Springer Theses

Recognizing Outstanding Ph.D. Research

Aims and Scope

The series "Springer Theses" brings together a selection of the very best Ph.D. theses from around the world and across the physical sciences. Nominated and endorsed by two recognized specialists, each published volume has been selected for its scientific excellence and the high impact of its contents for the pertinent field of research. For greater accessibility to non-specialists, the published versions include an extended introduction, as well as a foreword by the student's supervisor explaining the special relevance of the work for the field. As a whole, the series will provide a valuable resource both for newcomers to the research fields described, and for other scientists seeking detailed background information on special questions. Finally, it provides an accredited documentation of the valuable contributions made by today's younger generation of scientists.

Theses may be nominated for publication in this series by heads of department at internationally leading universities or institutes and should fulfill all of the following criteria

- They must be written in good English.
- The topic should fall within the confines of Chemistry, Physics, Earth Sciences, Engineering and related interdisciplinary fields such as Materials, Nanoscience, Chemical Engineering, Complex Systems and Biophysics.
- The work reported in the thesis must represent a significant scientific advance.
- If the thesis includes previously published material, permission to reproduce this must be gained from the respective copyright holder (a maximum 30% of the thesis should be a verbatim reproduction from the author's previous publications).
- They must have been examined and passed during the 12 months prior to nomination.
- Each thesis should include a foreword by the supervisor outlining the significance of its content.
- The theses should have a clearly defined structure including an introduction accessible to new PhD students and scientists not expert in the relevant field.

Indexed by zbMATH.

Yijing Huang

Towards the Optical Control of Resonantly Bonded Materials

An Ultrafast X-Ray Study

Doctoral Thesis accepted by Stanford University, CA, USA

Springer

Yijing Huang
University of Illinois
Champaign, IL, USA

ISSN 2190-5053 ISSN 2190-5061 (electronic)
Springer Theses
ISBN 978-3-031-42825-8 ISBN 978-3-031-42826-5 (eBook)
https://doi.org/10.1007/978-3-031-42826-5

This Springer imprint is published by the registered company Springer Nature Switzerland AG
The registered company address is: Gewerbestrasse 11, 6330 Cham, Switzerland

Paper in this product is recyclable.

Supervisor's Foreword

In her thesis, Dr. Yijing Huang describes key contributions to advancing the fundamental understanding of materials structure and dynamics from a microscopic perspective. In particular, she describes several advancements in time-domain methodologies using ultrafast pulses from x-ray free-electron lasers (FEL) to probe the interactions between electrons and phonons in photoexcited materials.

She focuses on one class of materials, the resonantly bonded materials. This class of materials hosts a number of structural phases that are sensitive to external parameters (e.g., temperature, pressure, chemical doping), and coupled to various functional properties (e.g., ferroelectricity, high thermoelectric figure of merit, a large change in optical constants upon crystallization and amorphization, also known as phase change materials). These materials thus serve as a playground for exploring new phases and new functionalities. In the thesis, she included studies on SnSe and Bi_2Te_3, which belong to this material family.

Using femtosecond time-resolved X-ray diffraction, she quantifies the coherent atomic motion trajectory upon sudden excitation of carriers in SnSe. This allowed her to reconstruct the nonequilibrium lattice structure and identify a novel lattice instability toward a higher-symmetry structure not found in equilibrium. She then investigates the excited-state phonon dispersion in SnSe using time-resolved X-ray diffuse scattering and gains important insight into how photoexcitation alters the strength of specific bonds leading to the novel lattice instability observed in X-ray diffraction. Finally, by combining ultrafast X-ray diffraction and ARPES, she performs quantitative measurements of electron-phonon coupling in Bi_2Te_3 and Bi_2Se_3.

The findings highlight the importance of time-resolved X-ray scattering techniques based on FELs, which reveal the details of the interplay between electron orbitals, atomic bonds, and structural instabilities. The microscopic information of electron-phonon interaction obtained from these methods can generally rationalize ways to control materials and design their functional properties.

Professor, Stanford University
Director, PULSE Institute
Stanford, CA, USA

David Reis

Acknowledgments

I want to thank the people who deeply impacted me during my Ph.D. years.

David Reis, my advisor, is a brilliant scientist with extensive scientific interests. As a scientist, he is always careful about making conclusions. He has non-compromising standards for the quality of science, which I appreciate (and more so toward the late years of my Ph.D.), and from which I benefited a lot. As an advisor, he is always willing to offer me his time and is very supportive of my scientific career. He took over a lot of administrative duties as PULSE director in my third year. Starting from my late fourth year, I began to ask for intense input, and he still managed to give me as much time as I requested. And when he is convinced of my results, he is always generous in giving me credit. He provided me with a healthy environment to grow as a young researcher so that I was never in hostile relations with colleagues and never felt over-pressured. Instead, I always feel respected and encouraged to pursue a full life.

Mariano Trigo, effectively my co-advisor. Beamtimes always have a higher success rate with him present. I can ask him about almost everything, from the big picture of physics to the nuts and bolts. If I had not built my research on the tools and knowledge he developed, everything would have been an order of magnitude more difficult. His efficiency, articulation, and balance between open-mindedness and prudence set a model for me years down the road.

Samuel Teitelbaum, the first postdoc that I worked with. Sam was passionate about everything, and he was always consciously providing mentorship. He taught me 80% percent of the fundamental knowledge I needed to kick-start the research. He was pedagogical and patient (we were gapped by a complete Ph.D.!), which I appreciate as a fresh beginner stumbling on English. He has a green thumb that makes people around him thrive. I would not have survived in the Reis group for the first three years if he were not there.

I would also like to thank other collaborators, Jonathan Sobota, Ivana Savic, Stephen Fahy, Shan Yang, and Olivier Delaire, for their insightful scientific discussions. Colleagues at Reis group, Viktor, Chance, Gilberto, Gal, Quynh, Ryan, and Johann, with whom I worked through beamtimes together.

As an international student, I also would like to thank the support from SLAC admins and Stanford Applied Physics admins: Paula, Patrice, Fiona, and Claire at the Applied Physics department and Michelle, Sheetal, Corrina, Margo, Chris, Mia, and many others at PULSE/SIMES, who sorted out a lot of tedious details for me throughout the past few years.

Finally, I want to thank the people who accompanied me on my Ph.D. journey and helped me build my identity: my fellows and mentors from the Christian Evangelical Fellowship, who gave me a home where I sought solace and spiritual support; friends who shared their lives with me, put their trust in me, and walked through the mountains high and valleys low with me; my parents, whose love is unmatched by anything else. I especially thank my mother, who suffered my sufferings and took them on as her burdens, yet she was strong enough to turn them into courage and cheerfulness.

My faith in physics was tested, destroyed, and rebuilt in the past six years. I rediscovered, eventually, that physics never cheats, and it never hides. God crafts the laws of physics. He also makes creatures with hearts that crave eternity and minds that discern the unchanging laws so that when they discover the laws He founds, their hearts could feel His joy.

May all glory be to God. He wipes away all the tears. His promises never fail.

For the publication of this book, I would like to give special thanks to my thesis committee members for the nomination, Alfred Zong who kindly shared his experiences of publishing in the Springer thesis series, and Springer editors Sam Harrison, Barbara Amorese, and Clement Wilson who helped me work toward the production.

Parts of This Thesis Have Been Published in the Following Journal Articles

Y. Huang, et al., Observation of a novel lattice instability in ultrafast photoexcited SnSe. Phys. Rev. X **12**, 011029 (2022)

Y. Huang, et al., Nonthermal bonding origin of a novel photoexcited lattice instability in SnSe. Phys. Rev. Lett. **131**, 156902 (2023)

Y. Huang, et al., Ultrafast measurements of mode-specific deformation potenials of Bi_2Te_3 and Bi_2Se_3. Phys. Rev. X **13**, 041050 (2023)

Contents

Chapter 1
Ultrafast X-Ray Scattering and Non-equilibrium States of Matter

Abstract This chapter offers a concise summary of recent studies conducted in the field of light-driven phases using ultrafast X-ray scattering. It aims to provide an overview of the research conducted in this area and to place the work presented in this dissertation within the broader context of ongoing investigations. In general, researchers employ ultrafast lasers to both generate and characterize non-equilibrium states of matter. X-ray scattering stands out as a particularly valuable tool in this regard. It offers unique advantages in revealing the emergent order of non-equilibrium states, accurately quantifying atomic motion, and probing microscopic interactions on atomic scales. These capabilities make X-ray scattering an indispensable technique for understanding the intricate dynamics of non-equilibrium systems.

This thesis is revised from the one submitted to Stanford University June 2022, in order to make the introduction sections more accessible.

The central focus of this thesis is the generation of non-equilibrium states of matter and their characterization through time-resolved spectroscopies. Time-domain studies offer more than just an alternative to frequency-domain spectroscopy, providing finer frequency resolution and easier access to low-energy excitations. The non-equilibrium states of matter created by ultrafast pump laser pulses constitute a significant and distinctive aspect of ultrafast spectroscopies.

Numerous intriguing natural processes, such as chemical reactions, photosynthesis, and catalysis, inherently occur in non-equilibrium conditions. Likewise, important applications in modern technologies, such as fast electronics and photovoltaics, also take place in non-equilibrium settings. The non-equilibrium state of matter induced by light holds the potential for on-demand materials engineering.

Utilizing ultrafast laser excitation, it becomes possible to modify the energy barriers between closely related material phases, thereby altering the delicate energetic equilibrium and potentially generating structures with novel functional properties that are inaccessible in thermal equilibrium [1]. In the past decade, researchers have made significant advancements in designing tailored pump sources to drive

Y. Huang, *Towards the Optical Control of Resonantly Bonded Materials*, Springer Theses, https://doi.org/10.1007/978-3-031-42826-5_1

non-equilibrium states [2–6], as well as developing more powerful microscopic probes [7–12].

My research not only focuses on creating innovative non-equilibrium states of matter but also aims to comprehend their formation and properties. In this chapter, I provide a concise overview of the ultrafast scattering technique, highlighting its unique advantages in characterizing light-induced non-equilibrium states.

I find it interesting to draw a comparison between the motivation for creating novel light-driven on-demand states in materials and the motivation for controlling chemical reaction pathways in molecular systems [13]. In the case of molecules, reactions can be affected by exploiting coherences in electronic and vibrational degrees of freedom, while in solids, the language of phase transitions and symmetry breaking is used, since there is long-range order in a lattice and the entire piece of material containing a large number of atoms on the order of 10^{23} must be considered as a single entity. Each of the three sections in this chapter centers around one aspect of this language: emergent order, uniform ($\mathbf{q} \rightarrow 0$) collective excitations of ordered systems, and dynamics of $\mathbf{q} \neq 0$. Each section aims to provide context for one of the three pieces of work (Chaps. 5–7) presented in this dissertation.

All experiments that form the main results of this thesis were made possible by recent advances in free-electron laser (FEL) sources [14–18], which allow for microscopic understanding of lattice dynamics [19–23].

1.1 Ultrafast Diffraction and Emergent Order

Diffraction techniques are useful for revealing new orders that form in materials. For instance, ultrafast X-ray diffraction of VO_2 has shown that photoexcitation can induce a sub-picosecond transformation from the monoclinic phase to a high-temperature rutile phase, which coincides with an insulator-to-metal transition [24]. As depicted in Fig. 1.1(a), the diffraction peak at low angles (indicated by the red arrow) that appears at 1ps (solid curve) corresponds to the rutile phase, which is absent before the pump (broken curve at -300 fs). The change in the diffraction peak feature over time is presented in Fig. 1.1(b), where the intensities are normalized by the pre-pump diffraction intensity. This monoclinic-to-rutile phase transition is the first-ever observed ultrafast solid-solid phase transition, which has since been studied using other time-resolved diffraction techniques [26–28].

In certain cases, large atomic motion triggered by the pump laser can result in metastable phases that do not exist in thermal equilibrium. For example, ultrafast electron diffraction experiments have demonstrated that a THz electric field can drive significant interlayer shear motion from a non-centrosymmetric T_d phase to a metastable centrosymmetric 1T' phase in WTe_2 at high fields. The T_d phase is the non-centrosymmetric ground-state structure, while the 1T' phase is the excited-state centrosymmetric structure. It has been suggested that a topological phase transition can be induced simultaneously with large atomic motions [29].

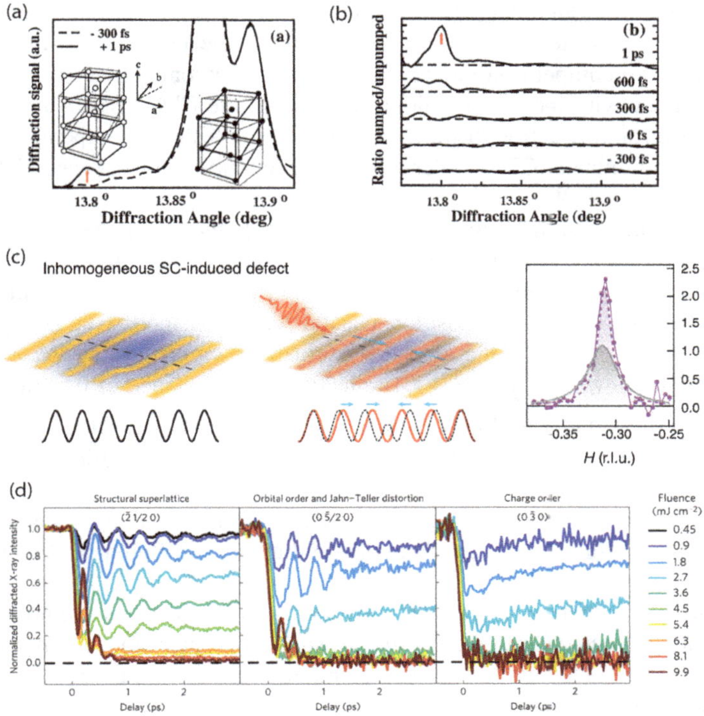

Fig. 1.1 (a–b) Time-resolved X-ray diffraction showing a solid-solid phase transition in VO_2. (a) Angle-dependent X-ray diffraction signal measured for negative and positive time delays. The dashed and continuous curves correspond to delays of −300 fs and 1 ps, respectively. The two crystalline structures, monoclinic insulating phase and tetragonal metallic phase, correspond to high and low diffraction angles, respectively. (b) Ratio between the X-ray diffraction of the excited and unpumped crystal. (c) X-ray resonant diffraction of $YBa_2Cu_3O_{6.67}$ charge density wave. The first plot is a schematic of the charge density wave in the ground state, with superconducting patches. The second plot represents the state after the quench of the superconducting state by a laser pulse. The third plot shows the simulated scattering profile for the equilibrium (gray solid lines) and photoexcited (purple dashed lines) cases. The dots are the experimental measurement of the charge density wave peak at around 2 ps. (d) X-ray resonant diffraction of $Pr_{0.5}Ca_{0.5}MnO_3$ on peaks of structural order, orbital order, and charge order. Figures (a–b) are adapted from ref. [24] with permission from the American Physical Society. Figure (c) is reproduced from ref. [8] with permission from the American Association for the Advancement of Science. Figure (d) is reproduced from ref. [25] with permission from Springer Nature

Ultrafast diffraction is a powerful tool not only for revealing emergent orders in non-equilibrium states but also for providing insights into how different orders compete with one another. For example, a recent ultrafast resonant diffraction study of the charge density wave (CDW) in $YBa_2Cu_3O_{6.67}$ showed that upon sudden quenching of the superconductivity by an infrared laser pulse, the correlation length of the CDW increased twice as compared to the equilibrium state, as evidenced by the change in the width of the CDW diffraction peak (gray curve for equilibrium and purple curve for excited states, in Fig. 1.1(c) right panel) [8]. This suggests that in equilibrium, the competition between CDW and superconductivity takes the form of superconductivity-induced topological defect in the CDW. In another example, ultrafast electron diffraction of the CDW in $TbTe_3$ revealed a competition between a strengthened uniaxial CDW order along a direction and a weakened CDW order along crystal axis c [30].

Moreover, ultrafast diffraction studies of ordered phases in materials can help us understand the nature of phase transitions. In $Pr_{0.5}Ca_{0.5}MnO_3$, for instance, despite the complex coupling between structural, orbital, and charge order in equilibrium, the dynamics of the phase transition can be well described with a single time-dependent order parameter that depends exclusively on electronic excitation [25]. Figure 1.1(d) shows that the $(\bar{2}\ \frac{1}{2}\ 0)$ reflection off resonance at 6.53 keV is only sensitive to the structural atomic motion, while the $(0\ \frac{\bar{5}}{2}\ 0)$ peak at 6.553 keV is mainly sensitive to the Jahn-Teller distortion. The $(0\ \bar{3}\ 0)$ reflection at 6.555 keV, which is resonantly enhanced, is mainly sensitive to charge order. By approximating the relevant atomic motions during the phase transition as four effective modes in a Ginzburg-Landau-like model, the coherent oscillations of different Bragg reflections can be well reproduced.

In solid-state materials, it is commonly assumed that electrons and holes quickly relax to a quasi-equilibrium electronic distribution after above-gap excitation, resulting in a loss of memory of the initial excitation [31–33]. As a result, the same order parameter that describes symmetry-breaking phase transitions in equilibrium is often applicable in photoexcited non-equilibrium states, provided that the temperature is replaced with electronic density or excitation fluence [25, 34, 35]. This seemingly limits the prospects of materials control. Therefore, there has been significant effort in exploring alternative ways to excite structural distortions in materials below the bandgap, such as terahertz and mid-infrared fields, in order to avoid heating the electronic degrees of freedom [36–42]. Although it is widely accepted that effective temperature models cannot be entirely accurate, their deviations from the photoexcited states in reality often appear too subtle or are invisible to conventional spectroscopies.

In my thesis, I use ultrafast X-ray diffraction to demonstrate that above-bandgap photoexcitation of SnSe results in a distinct structural instability that cannot be achieved by raising the temperature. The analysis is unambiguous due to a quantitative reconstruction of the sudden structural distortion following photoexcitation and the accompanying coherent phonon motion, including the *phase* of the sinusoidal trajectory. Only X-ray scattering probe based on FEL can achieve this quantitative reconstruction on the 10 ps timescale, and the association of this novel high symmetry structure with the results cannot be revealed with optical spectroscopies. These findings suggest that above-gap excitation offers previously unanticipated tunability and selectivity and a microscopic approach to materials control can be achieved by gaining a better understanding of the initial electronic excitations and probing of the coupled atomic motions.

1.2 Uniform Coherent Dynamics

In addition to studying phase transformations, diffraction probes can also investigate novel excitations and coherent dynamics in materials [41, 43]. In this section, I only talk about coherent dynamics of $\mathbf{q} \rightarrow 0$, which means the dynamics happen nearly uniformly for all unit cells in the crystal.

For instance, ultrafast X-ray diffraction has revealed a novel collective excitation called vortexon that is created by THz pulses in $[(PbTiO_3)_n/(SrTiO_3)_n]_m$ oxide superlattices, a type of polar vortices with much higher frequencies and smaller lateral sizes compared to magnetic vortices [43]. Figure 1.2(b) shows that the vortexon mode significantly softens at lower temperatures, indicating a condensation of the mode and a symmetry-breaking phase transition. Figure 1.2(c) shows the curl of the polarization field, and the maximum atomic displacement occurs at the largest curl of polarization.

Coherent dynamics of non-equilibrium states can reveal how energy is transferred between different degrees of freedom. In an experiment demonstrating the ultrafast Einstein-de Haas effect, Dornes et al. showed that over 80% of the angular momentum lost from spins in the transient demagnetization of ferromagnetic iron is transferred to the lattice in the form of a transverse strain wave, by measuring the scattering intensity along the truncation rod of the Bragg peak [44] ($\mathbf{q} \rightarrow 0$ limit). Figure 1.2(d, e) shows the dispersive modes along $\mathbf{q}_z$, which have the velocity of the sound wave. Another example is the non-equilibrium structure in $YBa_2Cu_3O_{6.5}$, as determined by femtosecond X-ray diffraction, revealing a transient change in the O–Cu distance that is simultaneous with the buildup of a finite superfluid density [38].

Studies of coherent dynamics using a combination of diffraction and electronic structure-sensitive probes such as angle-resolved photoemission spectroscopy (ARPES) can be powerful in quantifying couplings in the condensed phase. In a femtosecond measurement where a coherent phonon motion modulates both the diffraction intensity and the electronic band energy of FeSe, the deformation potential, a microscopic quantity, which is the amount of electron band shift per atomic displacement, can be quantified [45]. Figure 1.2(f–h) illustrates this. A combined measurement of time-resolved diffraction and ARPES can be useful in characterizing transient state properties, such as in $TiSe_2$, where 2D domain walls formed by optical excitation are accompanied by transient 2D ordered electronic states [46].

In the thesis, I use the femtosecond lock-in technique to characterize the surface state of the prototypical topological materials Bi_2Te_3 and Bi_2Se_3. I refine the methodology and obtain the deformation potential of the surface state and the A_{1g} mode, a fundamental quantity that can be useful for understanding surface state transport. The method can be used to probe topological phase transition coupled with structural changes, which is suggested in ref. [29].

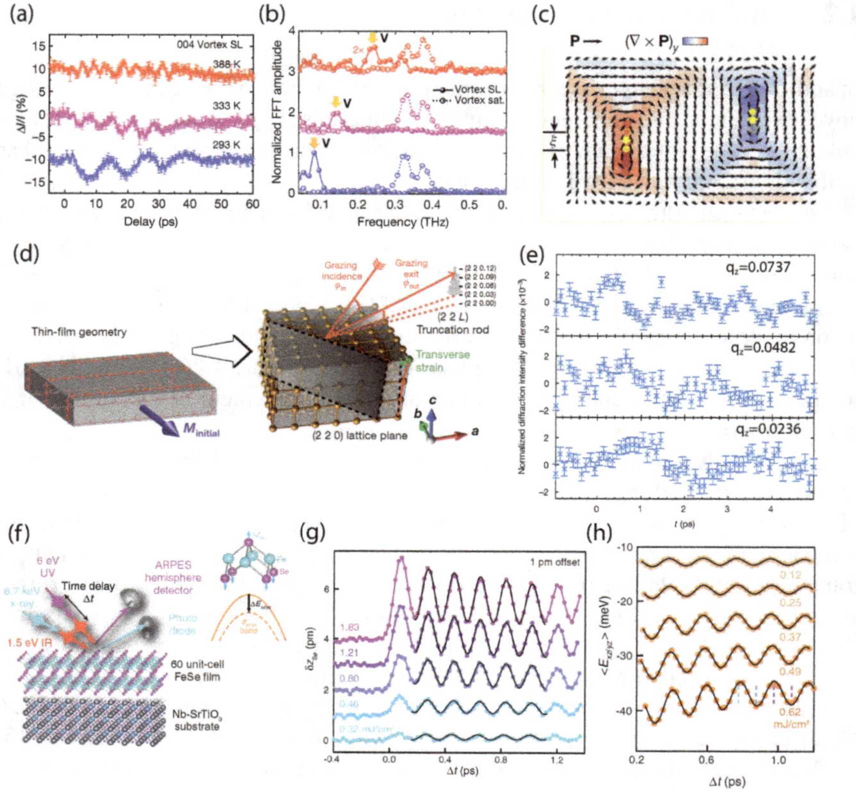

Fig. 1.2 (a–c) $[(PbTiO_3)_n/(SrTiO_3)_n]_m$ oxide superlattices. (a) Normalized change of diffraction intensity of superlattice (SL) peak of vortex structures as a function of delay at various temperatures. (b) Normalized Fourier spectra of the intensity oscillations of the 004 SL peak (solid). The orange arrows indicate the tunable vortexon mode. (c) Calculated curl of polarization (P) at the time of sinusoidal maximum (quarter of the oscillation period). The yellow arrows indicate the shift of vortex cores (solid dots) from the equilibrium positions (open yellow circles). (d–e) Ultrafast Einstein-De Haas effect in ferromagnetic iron. (d) Ultrafast Einstein-De Haas effect in a thin-film sample (left) and the geometry for X-ray diffraction experiment (right). The left panel is a schematic of the experimental result; the red arrows indicate transverse forces on the surface, which arise from the transfer of angular momentum. In the right panel, the diffraction along the truncation rod shows a transverse strain is launched by transferred angular momentum from demagnetization. (e) The experimental difference between the diffraction intensities measured in antiparallel magnetization directions, normalized to the sum of these intensities. The quantity shown here is the net response of the transverse strain wave to the demagnetization. (f–h) Time-resolved X-ray diffraction and ARPES on $FeSe/SrTiO_3$. (f) Experimental setup. Inset is a schematic of the A_{1g} phonon mode (top), which periodically modulates the electronic band energies (bottom). (g) Displacement of atom Se, as calculated from the diffraction data. (h) Momentum-averaged energy shifts of the d_{xz}/d_{yz} bands. Figures (a–b) are reproduced from ref. [43] with permission from Springer Nature. Figures (d–e) are reproduced from ref. [44] with permission from Springer Nature. Figures (f–h) are reproduced from ref. [45] with permission from the American Association for the Advancement of Science

1.3 Going Beyond the Zone Center

By going beyond the zone center ($\mathbf{q} \neq \mathbf{0}$) in a scattering experiment, we can access much more information on structural dynamics, such as the visualization of the polarons [47], domain walls [35], and potentially other topological defects as well as the related critical behavior [48–50].

Probing the large area of diffuse scattering in reciprocal space allows one to identify how a phase transition occurs. In a diffraction experiment that only measures the uniform behavior of unit cells, it is impossible to distinguish between a displacive-type and an order-disorder-type phase transition, as shown in Fig. 1.3(c). However, diffuse scattering measurements can provide this information [27]. For instance, in the photoinduced phase transition of VO_2 from the monoclinic phase to the rutile phase, the system explores a large volume of phase space within the short timescale of the transition, as shown in Fig. 1.3(d). Furthermore, these measurements can help us understand the effect of inertial dynamics on the speed of phase transitions.

Time-resolved diffuse scattering also enables the reconstruction of excited-state phonon dispersions, which is a time-domain analogue of inelastic scattering measurements. Trigo et al. first demonstrated this technique, called Fourier transform inelastic X-ray scattering [51], which allows for the extraction of interatomic forces in the excited states [53] and direct measurements of parametric phonon downconversion [52], among other applications [54, 55]. Figure 1.3(c–d) illustrates the working principle of this technique, and the movies of the diffuse scattering pattern with ripples provide a time-domain illustration of the dispersive modes. By selecting a path in reciprocal space and mapping out the dispersion upon a Fourier transform of the time-resolved scattering signal from individual $\mathbf{q}$ points, the phonon dispersion can be determined. Figure 1.3(e–f) shows the phonon downconversion process in bismuth [52], where Fig. 1.3(e) is the schematic and Fig. 1.3(f) is the experimental data showing the buildup of the parametrically amplified mode (orange) with a $\frac{\pi}{2}$ phase shift compared to the downconverting A_{1g} phonon (blue).

As lattice dynamics measurements are important for understanding how materials transform between different structures, I propose that performing $\mathbf{q}$-resolved lattice dynamics measurements after photoexcitation can directly reveal the effect of valence charge redistribution on interatomic interactions, structural instabilities, and the vibrational modes of the lattice. By fitting an interatomic force model to the excited-state dispersion in SnSe, I determined that the photoinduced lattice instability is primarily due to changes in the fourth-nearest neighbor bonds that connect bilayers, with relatively little change to the intralayer resonant bonds. These measurements provide critical insight into the nonthermal bonding origin of the instability in SnSe and can establish the relationship between the uniform ($\mathbf{q} = 0$) structural distortions, the atomic bonding inferred from the ($\mathbf{q} \neq 0$) transient excited-state phonon dispersion, and the electron orbitals perturbed in the photoexcited states. Previously, only macroscopic properties related to electron response functions (e.g., Born effective charge, optical constants) in the photoexcited states

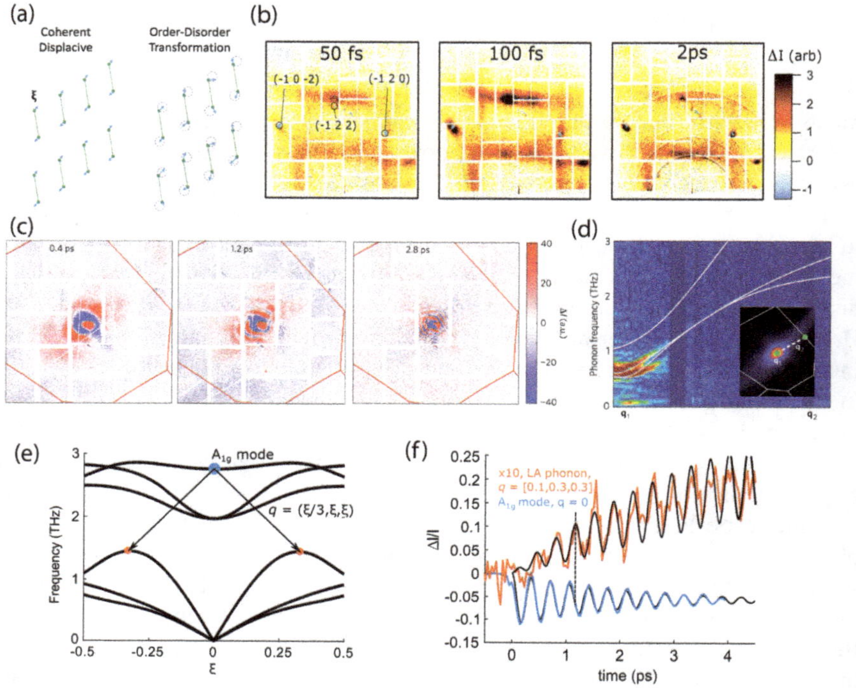

Fig. 1.3 (a–b) VO_2 ultrafast diffuse scattering. (a) Possible transition mechanisms between the monoclinic phase and the rutile phase upon photoexcitation. (b) Snapshots of the increase in the diffuse intensity at several delays after laser excitation. (c–d) Germanium ultrafast diffuse scattering. (c) Representative frames of the normalized intensity, sampled at different time points. (d) Acoustic dispersion along the section shown with dashed line in the inset. White lines are the calculated dispersions. (e–f) Parametric phonon downconversion in bismuth. (e) Phonon dispersion in bismuth, illustrating a downconversion of a A_{1g} phonon into two transverse acoustic phonons at $\mathbf{q}$ and $-\mathbf{q}$. (f) Experimental evidence of the decay process. The lower (blue) curve shows the relative intensity change of the (2 3 2) Bragg peak due to the decaying A_{1g} mode. The upper (orange) curve shows the relative intensity ($\times 10$) change of the diffuse scattering due to the amplified TA modes in a region near q = (0.1 0.3 0.3) in the (0 1 1) zone. Black lines are simulations. Figures (a–b) are reproduced from ref. [27] with permission from the American Association for the Advancement of Science. Figures (c–d) are reproduced from ref. [51] with permission from Springer Nature. Figures (e–f) are reproduced from ref. [52] with permission from the American Physical Society

measured at $\mathbf{q} = 0$ or microscopic $\mathbf{q} \neq 0$ properties in the equilibrium states could partially indicate this relationship. These new measurements bring us one step closer to understanding and controlling materials' properties under non-equilibrium conditions.

References

1. D.N. Basov, R.D. Averitt, D. Hsieh, Towards properties on demand in quantum materials. Nat. Mater. **16**(11), 1077–1088 (2017)
2. T. Feurer, J.C. Vaughan, K.A. Nelson. Spatiotemporal Coherent control of lattice vibrational waves. Science **299**(5605), 374–377 (2003)
3. K.-L. Yeh, M.C. Hoffmann, J. Hebling, K.A. Nelson. Efficient terahertz generation by optical rectification at 1035 nm. Opt. Express **15**(18), 11706–11713 (2007)
4. H. Bromberger, M. Foerst, T. Gebert, A. Cavalleri, B. Liu, A. Cartella, Generation of narrowband, high-intensity, carrier-envelope phase-stable pulses tunable between 4 and 18 THz. Opt. Lett. **42**(1), 129–131 (2017)
5. M. Budden, T. Gebert, M. Buzzi, G. Jotzu, E. Wang, T. Matsuyama, G. Meier, Y. Laplace, D. Pontiroli, M. Ricco, F. Schlawin, D. Jaksch, A. Cavalleri, Tuning metastable light-induced superconductivity in K3C60 with a hybrid CO2-Ti:Sapphire laser, in *Conference on Lasers and Electro-Optics (2021)* (2021), p. FF1A.5
6. J.G. Horstmann, H. Boeckmann, B. Wit, F. Kurtz, G. Storeck, C. Ropers, Coherent control of a surface structural phase transition. Nature **583**(7815), 232–236 (2020)
7. D.M. Fritz, D.A. Reis, B. Adams, R.A. Akre, J. Arthur, C. Blome, P.H. Bucksbaum, A.L. Cavalieri, S. Engemann, S. Fahy, R.W. Falcone, P.H. Fuoss, K.J. Gaffney, M.J. George, J. Hajdu, M.P. Hertlein, P.B. Hillyard, M. Horn-von Hoegen, M. Kammler, J. Kaspar, R. Kienberger, P. Krejcik, S.H. Lee, A.M. Lindenberg, B. McFarland, D. Meyer, T. Montagne, É.D. Murray, A.J. Nelson, M. Nicoul, R. Pahl, J. Rudati, H. Schlarb, D.P. Siddons, K. Sokolowski-Tinten, T. Tschentscher, D. von der Linde, J.B. Hastings, Ultrafast bond softening in bismuth: Mapping a solid's interatomic potential with X-rays. Science **315**(5812), 633–636 (2007)
8. S. Wandel, F. Boschini, E.H. da Silva Neto, L. Shen, M.X. Na, S. Zohar, Y. Wang, S.B. Welch, M.H. Seaberg, J.D. Koralek, G.L. Dakovski, W. Hettel, M.F. Lin, S.P. Moeller, W.F. Schlotter, A.H. Reid, M.P. Minitti, T. Boyle, F. He, R. Sutarto, R. Liang, D. Bonn, W. Hardy, R.A. Kaindl, D.G. Hawthorn, J.S. Lee, A.F. Kemper, A. Damascelli, C. Giannetti, J.J. Turner, G. Coslovich, Enhanced charge density wave coherence in a light-quenched, high-temperature superconductor. Science **376**(6595), 860–864 (2022)
9. A.K. Mills, S. Zhdanovich, M.X. Na, F. Boschini, E. Razzoli, M. Michiardi, A. Sheyerman, M. Schneider, T.J. Hammond, V. Süss, C. Felser, A. Damascelli, D.J. Jones, Cavity-enhanced high harmonic generation for extreme ultraviolet time- and angle-resolved photoemission spectroscopy. Rev. Sci. Instrum. **90**, 083001 (2019)
10. T. Danz, T. Domroese, C. Ropers, Ultrafast nanoimaging of the order parameter in a structural phase transition. Science **371**(6527), 371–374 (2021)
11. T.E. Glover, D.M. Fritz, M. Cammarata, T.K. Allison, S. Coh, J.M. Feldkamp, H. Lemke, D. Zhu, Y. Feng, R.N. Coffee, M. Fuchs, S. Ghimire, J. Chen, S. Shwartz, D.A. Reis, S.E. Harris, J.B. Hastings. X-ray and optical wave mixing. Nature **488**(7413), 603–608 (2012)
12. M. Schultze, K. Ramasesha, C.D. Pemmaraju, S.A. Sato, D. Whitmore, A. Gandman, J.S. Prell, L.J. Borja, D. Prendergast, K. Yabana, D.M. Neumark, S.R. Leone, Attosecond bandgap dynamics in silicon. Science **346**(6215), 1348–1352 (2014)
13. P. Nuernberger, G. Vogt, T. Brixner, G. Gerber, Femtosecond quantum control of molecular dynamics in the condensed phase. Phys. Chem. Chem. Phys. **9**, 2470–2497 (2007)
14. P. Emma, R. Akre, J. Arthur, R. Bionta, C. Bostedt, J. Bozek, A. Brachmann, P. Bucksbaum, R. Coffee, F.-J. Decker, et al., First lasing and operation of an ångstrom-wavelength free-electron laser. Nat. Photon. **4**(9), 641 (2010)
15. T. Ishikawa, H. Aoyagi, T. Asaka, Y. Asano, N. Azumi, T. Bizen, H. Ego, K. Fukami, T. Fukui, Y. Furukawa et al., A compact X-ray free-electron laser emitting in the sub-Angstrom region. Nat. Photon. **6**(8), 540–544 (2012)
16. S. Abeghyan, M. Bagha-Shanjani, G. Chen, U. Englisch, S. Karabekyan, Y. Li, F. Preisskorn, F. Wolff-Fabris, M. Wuenschel, M. Yakopov, J. Pflueger, First operation of the SASE1 undulator system of the European X-ray free-electron laser. J. Synchrot. Radiat. **26**(2), 302–310 (2019)

17. H.-S. Kang, C.-K. Min, H. Heo, C. Kim, H. Yang, G. Kim, I. Nam, S.Y. Baek, H.-J. Choi, G. Mun, et al., Hard X-ray free-electron laser with femtosecond-scale timing jitter. Nat. Photon. **11**(11), 708–713 (2017)
18. C.J. Milne, T. Schietinger, M. Aiba, A. Alarcon, J. Alex, A. Anghel, V. Arsov, C. Beard, P. Beaud, S. Bettoni, et al., SwissFEL: the swiss X-ray free electron laser. Appl. Sci. **7**(7), 720 (2017)
19. S. Wall, M. Trigo, Recent developments in ultrafast X-ray techniques for materials science applications. Synchrot. Radiat. News **29**(5), 13–18 (2016)
20. A.M. Lindenberg, S.L. Johnson, D.A. Reis, Visualization of atomic-scale motions in materials via femtosecond X-ray scattering techniques. Ann. Rev. Mater. Res. **47**(1), 425–449 (2017)
21. M. Buzzi, M. Foerst, R. Mankowsky, A. Cavalleri, Probing dynamics in quantum materials with femtosecond X-rays. Nat. Rev. Mater. **3**(9), 299–311 (2018)
22. M. Dunne, X-ray free-electron lasers light up materials science. Nat. Rev. Mater. **3**(9), 290–292 (2018)
23. Y. Cao, D.G. Mazzone, D. Meyers, J.P. Hill, X. Liu, S. Wall, M.P.M. Dean, Ultrafast dynamics of spin and orbital correlations in quantum materials: an energy- and momentum-resolved perspective. Philosoph. Trans. Roy. Soc. A Math. Phys. Eng. Sci. **377**(2145), 20170480 (2019)
24. A. Cavalleri, C. Tóth, C.W. Siders, J.A. Squier, F. Ráksi, P. Forget, J.C. Kieffer, Femtosecond structural dynamics in vo 2 during an ultrafast solid-solid phase transition. Phys. Rev. Lett. **87**(23), 237401 (2001)
25. P. Beaud, A. Caviezel, S.O. Mariager, L. Rettig, G. Ingold, C. Dornes, S.W. Huang, J.A. Johnson, M. Radovic, T. Huber, T. Kubacka, A. Ferrer, H.T. Lemke, M. Chollet, D. Zhu, J.M. Glownia, M. Sikorski, A. Robert, H. Wadati, M. Nakamura, M. Kawasaki, Y. Tokura, S.L. Johnson, U. Staub, A time-dependent order parameter for ultrafast photoinduced phase transitions. Nat. Mater. **13**(10), 923–927 (2014)
26. V.R. Morrison, R.P. Chatelain, K.L. Tiwari, A. Hendaoui, Andrew Bruhács, M. Chaker, B.J. Siwick, A photoinduced metal-like phase of monoclinic VO2 revealed by ultrafast electron diffraction. Science **346**(6208), 445–448 (2014)
27. S. Wall, S. Yang, L. Vidas, M. Chollet, J.M. Glownia, M. Kozina, T. Katayama, T. Henighan, M. Jiang, T.A. Miller, D.A. Reis, L.A. Boatner, O. Delaire, M. Trigo, Ultrafast disordering of vanadium dimers in photoexcited VO2. Science **362**(6414), 572–576 (2018)
28. M. de la Peña, A. Gilberto, A.A. Correa, S. Yang, O. Delaire, Y. Huang, A.S. Johnson, T. Katayama, V. Krapivin, E. Pastor, D.A. Reis, S. Teitelbaum, L. Vidas, S. Wall, M. Trigo, Ultrafast lattice disordering can be accelerated by electronic collisional forces. Nat. Phys. **19**, 1489–1494 (2023)
29. E.J. Sie, C.M. Nyby, C.D. Pemmaraju, S.J. Park, X. Shen, J. Yang, M.C. Hoffmann, B.K. Ofori-Okai, R. Li, A.H. Reid, et al., An ultrafast symmetry switch in a weyl semimetal. Nature **565**(7737), 61–66 (2019)
30. A. Kogar, A. Zong, P.E. Dolgirev, X. Shen, J. Straquadine, Y.-Q. Bie, X. Wang, T. Rohwer, I.-C. Tung, Y. Yang, et al., Light-induced charge density wave in $LaTe_3$. Nat. Phys. **16**(2), 159–163 (2020)
31. S.I. Anisimov, B.L. Kapeliovich, T.L. Perelman et al., Electron emission from metal surfaces exposed to ultrashort laser pulses. Zh. Eksp. Teor. Fiz **66**(2), 776–781 (1974)
32. T.Q. Qiu, C.L. Tien, Heat transfer mechanisms during short-pulse laser heating of metals. J. Heat Transf. **115**(4), 835–841 (1993)
33. T. Shin, S.W. Teitelbaum, J. Wolfson, M. Kandyla, K.A. Nelson, Extended two-temperature model for ultrafast thermal response of band gap materials upon impulsive optical excitation. J. Chem. Phys. **143**(19), 194705 (2015)
34. T. Huber, S.O. Mariager, A. Ferrer, H. Schäfer, J.A. Johnson, S. Grübel, A. Lübcke, L. Huber, T. Kubacka, C. Dornes, C. Laulhe, S. Ravy, G. Ingold, P. Beaud, J. Demsar, S.L. Johnson, Coherent structural dynamics of a prototypical charge-density-wave-to-metal transition. Phys. Rev. Lett. **113**, 026401 (2014)

35. M. Trigo, P. Giraldo-Gallo, J.N. Clark, M.E. Kozina, T. Henighan, M.P. Jiang, M. Chollet, I.R. Fisher, J.M. Glownia, T. Katayama, P.S. Kirchmann, D. Leuenberger, H. Liu, D.A. Reis, Z.X. Shen, D. Zhu, Ultrafast formation of domain walls of a charge density wave in $SmTe_3$. Phys. Rev. **103**, 054109 (2021)
36. M. Rini, N. Dean, J. Itatani, Y. Tomioka, Y. Tokura, R.W. Schoenlein, A. Cavalleri et al., Control of the electronic phase of a manganite by mode-selective vibrational excitation. Nature **449**(7158), 72–74 (2007)
37. D. Fausti, R. Tobey, N. Dean, S. Kaiser, A. Dienst, M.C. Hoffmann, S. Pyon, T. Takayama, H. Takagi, A. Cavalleri, Light-induced superconductivity in a stripe-ordered cuprate. Science **331**(6014), 189–191 (2011)
38. R. Mankowsky, A. Subedi, M. Först, S.O. Mariager, M. Chollet, H.T. Lemke, J.S. Robinson, J.M. Glownia, M.P. Minitti, A. Frano et al., Nonlinear lattice dynamics as a basis for enhanced superconductivity in $YBa_2Cu_3O_{6.5}$. Nature **516**(7529), 71–73 (2014)
39. X. Li, T. Qiu, J. Zhang, E. Baldini, L. Jian, A.M. Rappe, K.A. Nelson, Terahertz field induced ferroelectricity in quantum paraelectric SrTiO3. Science **364**(6445), 1079–1082 (2019)
40. M. Kozina, M. Fechner, P. Marsik, T. van Driel, J.M. Glownia, C. Bernhard, M. Radovic, D. Zhu, S. Bonetti, U. Staub et al., Terahertz-driven phonon upconversion in SrTiO 3. Nat. Phys. **15**(4), 387–392 (2019)
41. T. Kubacka, J.A. Johnson, M.C. Hoffmann, C. Vicario, S. de Jong, P. Beaud, S. Grübel, S.-W. Huang, L. Huber, L. Patthey, Y.-D. Chuang, J.J. Turner, G.L. Dakovski, W.-S. Lee, M.P. Minitti, W. Schlotter, R.G. Moore, C.P. Hauri, S.M. Koohpayeh, V. Scagnoli, G. Ingold, S.L. Johnson, U. Staub, Large-amplitude spin dynamics driven by a THz pulse in resonance with an electromagnon. Science **343**(6177), 1333–1336 (2014)
42. T. Kampfrath, K. Tanaka, K.A. Nelson, Resonant and nonresonant control over matter and light by intense terahertz transients. Nat. Photon. **7**(9), 680–690 (2013)
43. Q. Li, V.A. Stoica, M. Paściak, Y. Zhu, Y. Yuan, T. Yang, M.R. McCarter, S. Das, A.K. Yadav, S. Park, C. Dai, H.J. Lee, Y. Ahn, S.D. Marks, S. Yu, C. Kadlec, T. Sato, M.C. Hoffmann, M. Chollet, M.E. Kozina, S. Nelson, D. Zhu, D.A. Walko, A.M. Lindenberg, P.G. Evans, L.Q. Chen, R. Ramesh, L.W. Martin, V. Gopalan, J.W. Freeland, J. Hlinka, H. Wen, Subterahertz collective dynamics of polar vortices. Nature **592**(7854), 376–380 (2021)
44. C. Dornes, Y. Acremann, M. Savoini, M. Kubli, M.J. Neugebauer, E. Abreu, L. Huber, G. Lantz, C.A.F. Vaz, H. Lemke, E.M. Bothschafter, M. Porer, V. Esposito, L. Rettig, M. Buzzi, A. Alberca, Y.W. Windsor, P. Beaud, U. Staub, D. Zhu, S. Song, J.M. Glownia, S.L. Johnson, The ultrafast Einstein'96de Haas effect. Nature **565**(7738), 209–212 (2019)
45. S. Gerber, S.-L. Yang, D. Zhu, H. Soifer, J.A. Sobota, S. Rebec, J.J. Lee, T. Jia, B. Moritz, C. Jia et al., Femtosecond electron-phonon lock-in by photoemission and X-ray free-electron laser. Science **357**(6346), 71–75 (2017)
46. S. Duan, Y. Cheng, W. Xia, Y. Yang, X. Chengyang, F. Qi, C. Huang, T. Tang, Y. Guo, W. Luo, D. Qian, D. Xiang, J. Zhang, W. Zhang, Optical manipulation of electronic dimensionality in a quantum material. Nature **595**(7866), 239–244 (2021)
47. B. Guzelturk, T. Winkler, T.W.J. Van de Goor, M.D. Smith, S.A. Bourelle, S. Feldmann, M. Trigo, S.W. Teitelbaum, H.G. Steinrück, G.A. de la Pena, R. Alonso-Mori, D. Zhu, T. Sato, H.I. Karunadasa, M.F. Toney, F. Deschler, A.M. Lindenberg, Visualization of dynamic polaronic strain fields in hybrid lead halide perovskites. Nat. Mater. **20**(5), 618–623 (2021)
48. H. Toyoki, Structure factors of vector-order-parameter systems containing random topological defects. Phys. Rev. B **45**(5), 1965 (1992)
49. A.J. Bray, K. Humayun, Universal amplitudes of power-law tails in the asymptotic structure factor of systems with topological defects. Phys. Rev. E **47**(1), R9 (1993)
50. A.J. Bray, Topological defects and phase ordering dynamics, in *Formation and Interactions of Topological Defects* (Springer, Berlin, 1995), pp. 105–138

51. M. Trigo, M. Fuchs, J. Chen, M.P. Jiang, M. Cammarata, S. Fahy, D.M. Fritz, K. Gaffney, S. Ghimire, A. Higginbotham, S.L. Johnson, M.E. Kozina, J. Larsson, H. Lemke, A.M. Lindenberg, G. Ndabashimiye, F. Quirin, K. Sokolowski-Tinten, C. Uher, G. Wang, J.S. Wark, D. Zhu, D.A. Reis, Fourier-transform inelastic X-ray scattering from time- and momentum-dependent phonon'96phonon correlations. Nat. Phys. **9**(12), 790–794 (2013)
52. S.W. Teitelbaum, T. Henighan, Y. Huang, H. Liu, M.P. Jiang, D. Zhu, M. Chollet, T. Sato, É.D. Murray, S. Fahy, S. O'Mahony, T.P. Bailey, C. Uher, M. Trigo, D.A. Reis, Direct measurement of anharmonic decay channels of a coherent phonon. Phys. Rev. Lett. **121**, 125901 (2018)
53. S.W. Teitelbaum, T.C. Henighan, H. Liu, M.P. Jiang, D. Zhu, M. Chollet, T. Sato, É.D. Murray, S. Fahy, S. O'Mahony, T.P. Bailey, C. Uher, M. Trigo, D.A. Reis, Measurements of nonequilibrium interatomic forces using time-domain X-ray scattering. Phys. Rev. B **103**, L180101 (2021)
54. M.P. Jiang, M. Trigo, I. Savić, S. Fahy, É.D. Murray, C. Bray, J. Clark, T. Henighan, M. Kozina, M. Chollet et al., The origin of incipient ferroelectricity in lead telluride. Nat. Commun. **7**(1), 12291 (2016)
55. V. Krapivin, M. Gu, D. Hickox-Young, S.W. Teitelbaum, Y. Huang, G. De La Pe na, D. Zhu, N. Sirica, M.C. Lee, R.P. Prasankumar, A.A. Maznev, K.A. Nelson, M. Chollet, J.M. Rondinelli, D.A. Reis, M. Trigo, Ultrafast suppression of the ferroelectric instability in KTaO3. Phys. Rev. Lett. **129**(12), 1–6 (2022)

Chapter 2
Lattice Dynamics: Excitation and Probe

Abstract Lattice dynamics exhibit distinct characteristics in both thermal equilibrium and non-equilibrium states. Despite the differences, the fundamental concepts of phonons and electron-lattice interaction remain relevant in both scenarios. This chapter addresses both the ultrafast excitation mechanisms of Raman-active phonons and the light-matter interactions that facilitate the investigation of lattice dynamics using X-ray, under non-equilibrium conditions.

Structural instability is a key concept explored in this thesis. To lay the foundation for understanding bulk crystals, I will first present a chemist's perspective on structural instability, focusing on the role of chemical bonding. This approach offers a straightforward and valuable framework for comprehending the phenomenology of structural phase transitions. When a crystal adopts a stabilized phase, we can delve into the discussion of elementary excitations in the ordered structure. These include phonons, which represent the quanta of lattice vibrations and play a significant role in characterizing the crystal's behavior.

I first present the formulation of the dynamical matrix and phonons as its eigenvectors or normal modes. We then look into how the interatomic interactions break down into ion-ion interaction without considering valence electrons (an ion refers to the combination of core electrons and nuclei), and interatomic interactions due to a valence charge redistribution ensuing movement of ions. The latter can thus be formulated with an electron response function in the perturbative regime. The electron-phonon interaction gives rise to key features in both electron and phonon band structures, such as the condensation of electrons into ordered charged density waves and Kohn anomalies in phonon dispersions. It also determines the transport properties of both electrons and phonons. Electron-phonon interactions can be classified based on whether there is long-range electric field in the crystal. In Chap. 7, we will quantify the deformation potential, which describes the short-range interaction between electrons and phonons. In equilibrium, the electron phonon interaction determines the scattering cross section of a Raman scattering process. Raman spectroscopy can be a powerful tool for studying phonon dynamics near the zone center.

Y. Huang, *Towards the Optical Control of Resonantly Bonded Materials*, Springer Theses, https://doi.org/10.1007/978-3-031-42826-5_2

In typical ultrafast spectroscopy experiments, two light pulses are used: the pump pulse, which induces dynamics in the material, and the probe pulse, which detects and characterizes those dynamics. These pulses are spatially overlapped to ensure their interaction in one region of the sample. The choice of pump and probe pulses can be tailored independently based on the specific research objectives. Different combinations of pump and probe wavelengths, energies, and time delays can be utilized to address different aspects of the studied system. The exitation of coherent Raman active phonons occurs through a Raman process, where the incident light induce electron fluctuations which subsequently couple to the phonons. In this thesis, the focus is primarily on phonons excited through this Raman-like mechanism. Other mechanisms of phonon excitation are not extensively discussed. To probe the excited phonons, X-ray scattering techniques are employed in this thesis, providing the **q** (the momentum transfer) resolution of these collective excitations.

2.1 Structural Instabilities from a Chemical Bonding Perspective

The changes in chemical bonding that stabilize distorted structures have long been recognized in the field of coordination chemistry and are termed second-order Jahn-Teller effects. The energy of an electronic ground state $E(\delta)$ as a function of the coordinate of the distortion δ is [1]

$$E(\delta) = E_0 + \langle 0|(\partial H/\partial \delta)_0|0\rangle \delta + \frac{1}{2}\Bigg(\langle 0|(\partial^2 H/\partial^2 \delta)_0|0\rangle - 2\sum_n \frac{|\langle 0|(\partial H/\partial \delta)_0|n\rangle|^2}{E_n - E_0}\Bigg)\delta^2 \tag{2.1}$$

where higher-order terms are ignored. E_0 is the ground state energy of the undistorted structure, and E_n are the excited state energies. The linear δ term is nonzero only for degenerate states, and we shall focus on the second-order term in δ. The first part of the second-order term accounts for the elastic energy cost associated with the distortion, which is primarily influenced by the Coulomb repulsion between electron clouds. However, this term tends to be small for closed-shell ions without spatially extended valence electrons. On the other hand, the second part of the second-order term always reduces the energy. For this reduction to outweigh the energy cost term and induce a structural distortion, there must be low-lying excited states available (with small differences $E_n - E_0$). Additionally, the term $\langle 0|(\partial H/\partial \delta)_0|n\rangle$, which indicates the formation of new chemical bonds in a lower symmetry configuration, needs to be sufficiently significant. It's important to note that this description provides a local perspective and does not account for the periodicity of an extended crystal. In the case of a periodic crystal, a cooperative

Jahn-Teller effect becomes relevant. This effect refers to a phase transition driven by the interaction between the crystal lattice and localized electron orbitals. It involves both a splitting of electronic states and a uniform distortion of the lattice that lowers the symmetry. By drawing on the chemical bonding view of the Jahn-Teller distortion, valuable insights can be gained when considering structural phase transitions in crystals. The exploration of this idea will be contextualized within the research study presented in Chap. 5.

2.2 Lattice Dynamics

In the description of a solid, we can consider it as composed of nuclei that oscillate around their equilibrium positions, while electrons follow these movements. The electron population is divided into two categories: core electrons, which move rigidly with the nuclei, and valence electrons, whose motion is primarily determined by solving the many-body Schrödinger equation. However, in practice, the Born-Oppenheimer approximation or the adiabatic approximation is commonly employed. This approximation assume that the lighter valence electrons adjust adiabatically to the motion of the heavier ion. It is generally assumed that the electrons remain in their instantaneous ground state, although the validity of this assumption in the ultrafast timescale is yet to be confirmed.

Dynamical Matrix and Phonons

As an approximation, the vibrational dynamics of ions in a periodic solid can be thought to be harmonic, and we can formulate the dynamical matrix [2]. The crystal potential energy in the harmonic approximation is

$$U = U_0 + \frac{1}{2} \sum_{m,m;=1}^{N} \sum_{s,s'} \mathbf{u}_{m,s} \frac{\partial^2 U}{\partial \mathbf{r}_{ms} \partial \mathbf{r}_{m's'}} \mathbf{u}_{m's'} \tag{2.2}$$

where m is the lattice site $\mathbf{R}_m$, s is the atom at $\boldsymbol{\tau}_s$ inside the unit cell, $\mathbf{u}$ is the atomic displacement from equilibrium, $\mathbf{r}$ is the atomic equilibrium position, and $\mathbf{r}_{m,s} = \mathbf{R}_m + \boldsymbol{\tau}_s$. N is the total number of unit cells in the crystal. The interatomic force matrices in real space (each a 3×3 matrix) are

$$F_{msm's'} = -\frac{\partial^2 U}{\partial \mathbf{r}_{ms} \partial \mathbf{r}_{m's'}} \tag{2.3}$$

The equation of motions is

$$M_s \ddot{\mathbf{u}}_{m,s} = \sum_{m's'} F_{msm's'} \mathbf{u}_{m's'} \tag{2.4}$$

M_s is the atomic mass. If we assume

$$\mathbf{u}_{m,s} = \frac{1}{\sqrt{M_s}} \sum_{\mathbf{q},j} A_{\mathbf{q}} \boldsymbol{\epsilon}^s_{\mathbf{q},j} e^{i\mathbf{q}(\mathbf{R}_m+\boldsymbol{\tau}_s)-i\omega_{\mathbf{q}}t} \tag{2.5}$$

and that $\mathbf{u}_{m,s}$ and $\mathbf{u}_{m',s}$ are only different by a phase, then solving the equation of motion is recasted into an eigenvalue problem

$$D(\mathbf{q})\boldsymbol{\epsilon}_{\mathbf{q},j} = \omega^2_{\mathbf{q},j}\boldsymbol{\epsilon}_{\mathbf{q},j} \tag{2.6}$$

where j is the index of the eigenvalue, and

$$D_{ss'}(\mathbf{q}) = -\sum_{m'} \frac{F_{msm's'}}{\sqrt{M_s M_{s'}}} \exp[i\mathbf{q}\cdot(\mathbf{R}_{mm'} + \boldsymbol{\tau}_{ss'})] \tag{2.7}$$

$\mathbf{R}_{mm'} = \mathbf{R}_m - \mathbf{R}_{m'}$ and $\boldsymbol{\tau}_{ss'} = \boldsymbol{\tau}_s - \boldsymbol{\tau}_{s'}$. The phonon polarization, or the eigenvector of the dynamical matrix, is $\boldsymbol{\epsilon}_{\mathbf{q},j} = (\mathbf{e}_{\mathbf{q}js_1}, \mathbf{e}_{\mathbf{q}js_2}, ...)$. $D_{ss'}(\mathbf{q})$ is called a dynamical matrix, which is a force matrix in the spatial Fourier domain. Its eigenvectors satisfy the orthonormal relation $\sum_s(\boldsymbol{\epsilon}^s_{\mathbf{q},j})^* \cdot \boldsymbol{\epsilon}^s_{\mathbf{q},j'} = \delta_{j,j'}$ and the periodicity $\boldsymbol{\epsilon}_{\mathbf{q}+\mathbf{G},j,s} = \boldsymbol{\epsilon}_{\mathbf{q},js}\exp(-i\mathbf{G}\cdot\boldsymbol{\tau}_s)$.

The interatomic interactions obey the acoustic sum rule

$$F^{ij}_{m,s;m'=m,s'=s} = \sum_{m\neq m',s\neq s'} F^{ij}_{m,s;m',s'} \tag{2.8}$$

so that

$$\sum_s D^{ij}_{ss'}(\mathbf{q}=\mathbf{0}) = 0 \tag{2.9}$$

where i, j denote the Cartesian coordinates (x, y, z). This ensures the discrete translational symmetry of the lattice or equivalently that the acoustic modes approach zero frequency as $\mathbf{q} \to \mathbf{0}$. When one obtains the interatomic forces from first-principle calculations in a finite-size lattice, the acoustic sum rule needs to be imposed for the phonon dispersions to make physical sense.

In essence, a phonon is an elementary excitation in the quantum mechanical treatment of vibrations in a crystal lattice or the quantum unit of a crystal lattice vibration. And Eq. 2.5 can be recast into the form

$$\mathbf{u}_{m,s} = \frac{1}{\sqrt{NM_s}} \sum_{\mathbf{q},j} \sqrt{\frac{\hbar}{2\omega_{\mathbf{q},j}}} \boldsymbol{\epsilon}^s_{\mathbf{q},j} e^{i\mathbf{q}\cdot(\mathbf{R}_m+\boldsymbol{\tau}_s)-i\omega_{\mathbf{q}}t}(b_{\mathbf{q}j} + b^+_{-\mathbf{q}j}) \tag{2.10}$$

upon replacing the position operator with the creation and annihilation operators. The commutation relations of annihilation operator are $[b_{\mathbf{q}j}, b_{-\mathbf{q}'j'^{+}}] = \delta_{jj'}\delta_{\mathbf{q}\mathbf{q}'}$, $[b_{\mathbf{q}j}, b_{-\mathbf{q}'j'}] = 0$, $[b^{+}_{\mathbf{q}j}, b^{+}_{-\mathbf{q}'j'}] = 0$.

Electron Response Function

When talking about lattice dynamics, it is indispensable to talk about electron dynamics that contributes to the effective interactions between ions. We can show that with adiabatic approximation, for the dynamical matrix, electron-phonon interaction can be taken care of with the electron response function, which plays a central role in the microscopic theory of lattice dynamics, on a phenomenological level. The capability of electron response function to capture the real crystal behavior serves as a criterion for adopting a phenomelogical few-parameter model. In more complicated cases, one may want to resort to first principle calculations. Simple phenomenological models, which only contain a few parameters but explicitly take into account the valence electrons, have been used to fit the measured phonon spectra [3–7], though each with their own limitations.

In semiconductors, the charge distribution cannot be approximated by point charges or closed-shell ions, but one can still approximate the interactions between atoms in a semiconductor to consist of the two types of contributions that tend to cancel each other:

(1) The force on an ion due to Coulomb fields from other ion cores (the first of the second-order δ term in Eq. 2.1)
(2) The force on an ion due to the potential set by the valence electrons redistributed by the displacement of other ions (the second of the second-order δ term in Eq. 2.1)

The electron-ion interaction is Coulombic outside the ion core but can deviate from Coulombic inside the core due to the effective repulsion that arise from the orthogonality between valence states and core states. This effective potential is called pseudopotential, and we denote it as $v(\boldsymbol{r})$, which takes the form $v(\boldsymbol{r}) = Z\frac{e^2}{r}$ outside the core, where Z is the ion charge. We assume that when ionic motions incur a small electric potential $\delta v(\boldsymbol{x})$, the change in the electron density distribution is

$$\delta n(\boldsymbol{x}) = \int d\boldsymbol{x}'\chi(\boldsymbol{x}, \boldsymbol{x}')\delta v(\boldsymbol{x}') \tag{2.11}$$

where $\chi(\boldsymbol{x}, \boldsymbol{x}')$ is the electron density response function. Without considering the phonon effect, χ can be calculated simply based on the electronic states [8], taking into account the exchange correlation. The induced electron charge density $\delta n(\boldsymbol{x})$ in turn induces an additional energy term $\delta n(\boldsymbol{x})v(\boldsymbol{x})$. The harmonic energy is then

$$U(\boldsymbol{r}_{ms},\boldsymbol{r}_{m's'}) = Z_s Z_{s'} v(\boldsymbol{r}_{ms} - \boldsymbol{r}_{m's'}) + \int dx \int d\boldsymbol{x}' v_s(\boldsymbol{r}_{ms} - \boldsymbol{x})\chi(\boldsymbol{x},\boldsymbol{x}')v_{s'}(\boldsymbol{x}' - \boldsymbol{r}_{m's'}) \tag{2.12}$$

The effective charge induced tends to cancel out the electric field incurred by the ion motion. That is, the change in second term of Eq. (2.12) right hand side usually has an opposite sign to the change in the first term. The total effective charge density related to atom Z_s is

$$\delta\rho(\boldsymbol{x},\boldsymbol{r}_s) = \left[-Z_s e\frac{\partial\delta(\boldsymbol{x}-\boldsymbol{r}_s)}{\partial\boldsymbol{r}_s} + e\int d\boldsymbol{x}'\chi(\boldsymbol{x}',\boldsymbol{x})(-Z_s)\frac{\partial v(\boldsymbol{x}'-\boldsymbol{r}_s)}{\partial\boldsymbol{x}'}\right]\delta\boldsymbol{r}_s \tag{2.13}$$

The form of Eq. 2.13 suggests expressing quantities in the spatial Fourier space ($\boldsymbol{q}$ space). In crystals, one can define Born effective charge in the form of a tensor, as the coefficient of the proportionality between macroscopic polarization and the ion displacement, assuming no macroscopic electric field in the crystal, which means that the polarization field has to neutralize the induced effective charge.

$$\nabla\cdot P + \delta\rho = 0 \tag{2.14}$$

so

$$iq_\alpha Z_{\alpha\beta,q} = \int d\boldsymbol{x} e^{-i\boldsymbol{q}\cdot\boldsymbol{x}}\frac{\delta\rho(\boldsymbol{x},\boldsymbol{r}_s)}{\delta\boldsymbol{r}_s} \tag{2.15}$$

and $Z_{\alpha\beta,q}$ is the Born effective charge. A large Born effective charge is associated with large polarizability since both quantities are related to a small electron response function $\chi(\boldsymbol{x}',\boldsymbol{x})$ or a minor screening effect from electrons. In non-polar crystals, the Born effective charge can vanish due to symmetry.

The above linear response formalism assumes the adiabatic condition; the electron density response is calculated at zero frequency and does not take into account the phonon contributions. Considering a finite frequency electron response function and phonon effects will introduce the concept of phonon self-energy, which can be related to phonon-phonon interactions and anharmonicity.

Electron-Phonon Interaction

Now that we established, in the adiabatic approximation, that we can use the linear response function of electrons to describe the interatomic interactions (see Eq. 2.13), we will then think about how the valence electrons affect the interatomic interactions and then in turn the phonons or the so-called electron-phonon coupling.

The Hamiltonian of a solid composed of electrons and phonons is

$$\hat{H} = \sum_{\mathbf{q}} \hbar\Omega_{\mathbf{q}} b^{\dagger}_{\mathbf{q}} b_{\mathbf{q}} + \sum_{\mu\mathbf{k}} E_{\mu,\mathbf{k}} c^{\dagger}_{\mu,\mathbf{k}} c_{\mu,\mathbf{k}} + \sum_{\mu\nu,\mathbf{kq}} \Xi^{\mu\nu}_{\mathbf{k+q},\mathbf{k}} b_{\mathbf{q}} c^{\dagger}_{\mu,\mathbf{k+q}} c_{\nu,\mathbf{k}} \tag{2.16}$$

The term $\sum_{\mathbf{q}} \hbar\omega_{\mathbf{q}} b^{\dagger}_{\mathbf{q}} b_{\mathbf{q}}$ is the Hamiltonian for phonons, $\sum_{\mu\mathbf{k}} c^{\dagger}_{\mu,\mathbf{k}} c_{\mu,\mathbf{k}}$ is the Hamiltonian for Bloch electrons (μ represents an electron band, while $\mathbf{k}$ represents the wave vector of the electron), and $c^{\dagger}_{\mu,\mathbf{k}}$ and $c_{\mu,\mathbf{k}}$ satisfy anticommutation relation for electrons. The microscopic theory of electron-phonon interaction is an investigation of $\Xi_{\mathbf{k+q},\mathbf{k}}$, which describes the coupling between electron states of wave vector $\mathbf{k}+\mathbf{q}$ and $\mathbf{k}$, through a phonon wave vector $\mathbf{q}$.

In metals, the electric potential due to the displaced ion is completely screened by the conduction electrons, and the electron-phonon interaction is short-ranged. But in non-metallic crystals, such as semiconductors and ionic insulators, there can be unscreened charges that lead to a long-range electric field and thus long-range electron-phonon coupling.

Deformation potential is a form of short-ranged electron-phonon coupling that does not take into account the macroscopic electric field. This type of interaction exists for all crystals. The displacement of an ion can locally affect the electrons, and a redistribution of electrons or a renormalization of the electron band structure will lead to energy change of the system. At long wavelength limit ($\mathbf{q} \to 0$), the deformation potential for the acoustic phonon can be defined as a change in the electron energy upon a uniform strain field, and the deformation potential for the optical phonon can be defined as the change in the electron energy upon the internal distortion along the phonon coordinate.

In a solid, the proper assumption is that there is no free charge and no source for dielectric displacement.

$$\nabla \cdot (\boldsymbol{P} + \epsilon \boldsymbol{E}) = 0 \tag{2.17}$$

ϵ is the dielectric function which is related to the linear response function and the Fourier transform of the pseudopotential, $\epsilon(\mathbf{q}) = \delta(\mathbf{q}) - v(\mathbf{q})\chi(\mathbf{q})$. Therefore, in the polar crystal (Born effective charge does not vanish), there will be a macroscopic electric field created by a short-range variation in charge density located around the displaced nuclei and screened by the dielectric medium. A macroscopic electric field gives rise to longitudinal optic phonon (LO) and transverse optical phonon (TO) splitting. The splitting is due to that fact that for LO, the electric field is opposite to the LO-induced dipole field to keep the dielectric displacement zero, leading to a higher restorative force of LO. It gives rise to the well-known Fröhlich interaction, where the electrons interact with LO through the LO-generated macroscopic electric field. In general, the dipole-induced electric field is not uniform on the microscopic scale, which gives rise to gradient in dipoles and in turn the macroscopic quadrupole fields, and a multi-pole expansion is proper.

In polar materials, the nonzero quadrupole gives rise to both the piezoelectric coupling (long-range) and deformation potential type of coupling (short-range). The piezoelectric coupling is only significant at long wavelength (since it is a multi-pole expansion due to Coulomb interaction, the $\Xi_{\mathbf{k}+\mathbf{q},\mathbf{k}}$ term comes with $1/|\mathbf{q}|$), while the long-range electric field is screened by carriers. For optical phonons, the Fröhlich interaction generally dominates (and piezoelectric coupling is weak since it is a high-order expansion). For longitudinal acoustic phonon (LA), usually, the short-range deformation potential dominates at room temperature. For transverse acoustic phonon (TA) that does not scatter electrons through the deformation potential mechanism, the electron-phonon interaction happens mainly through long-range piezoeletric (piezoacoustic) coupling.

For non-polar crystals, the short-range deformation potential interaction mechanism dominates, for both optical and acoustic phonons. Note here that the definition of polar crystals is not based on whether the crystal space group breaks inversion symmetry but based on whether there is a finite Born effective charge.

As $\mathbf{q} \to 0$, the requirement of the analyticity of $\Xi_{\mathbf{k}+\mathbf{q},\mathbf{k}}$ equals the requirement of charge neutrality and results in acoustic sum rule. I refer the readers to refs. [9–11] for details.

Raman Scattering

Raman scattering is a useful way to probe lattice dynamics in equilibrium. Macroscopically, Raman scattering can be well formulated using the fluctuation-dissipation theorem and Raman tensors (which are measured and tabulated for materials). Phonons are phenomenologically regarded to change the polarizability and thus are coupled to light. From that point of view, the phonon field is considered another classical field, just as an electric field would be treated in the classical formalism of nonlinear optics. The microscopic theory of the Raman process involves third-order time-dependent perturbation theory; see Fig. 2.1(a). In this thesis, I do not discuss the excitation or probing of non-Raman-active phonons (e.g., IR-active phonons in a centrosymmetric crystal). In a Raman process, the light field directly couples to the electron states through dipole interaction (processes 1 and 3 in Fig. 2.1) but not directly to phonons—we need electron-phonon interaction (process 2 of Fig. 2.1) for the light to indirectly couple to the phonon.[1] The macroscopic Raman tensors can be associated with $\Xi_{\mathbf{k}+\mathbf{q},\mathbf{k}}$. Below, I outline the procedure for the derivation and refer readers to ref. [12] for details.

The Hamiltonian of a solid and the light field is

$$\hat{H} = \sum_{\lambda} \hbar\omega_{\lambda}\hat{a}_{\lambda}^{\dagger}\hat{a}_{\lambda} + \sum_{\mathbf{q}} \hbar\Omega_{\mathbf{q}} b_{\mathbf{q}}^{\dagger} b_{\mathbf{q}} + \sum_{\mu\mathbf{k}} E_{\mu,\mathbf{k}} c_{\mu,\mathbf{k}}^{\dagger} c_{\mu,\mathbf{k}} + \sum_{\mu\nu,\mathbf{kq}} \Xi_{\mathbf{k}+\mathbf{q},\mathbf{k}}^{\mu\nu} b_{\mathbf{q}} c_{\mu,\mathbf{k}+\mathbf{q}}^{\dagger} c_{\nu,\mathbf{k}} + \sum_{\mu\nu,\mathbf{kk}',\lambda\lambda'} \Pi_{\mu\mathbf{k}\nu\mathbf{k}'}^{\lambda\lambda'} \hat{a}_{\lambda'}^{\dagger} c_{\mu,\mathbf{k}}^{\dagger} c_{\nu,\mathbf{k}'} \hat{a}_{\lambda} \tag{2.18}$$

[1] Light can directly couple to IR-active phonons though, but that is not a Raman process.

We adopted dipole approximation (the lowest-order interaction between electrons and light), and $\Pi_{\mu\mathbf{k}\nu\mathbf{k}'}$ represents the dipole interaction. The operators $\hat{a}_\lambda$ and $\hat{a}_\lambda^\dagger$ are the annihilation and creation operators of a photon in the mode λ which denotes the polarization and photon wave vector. They satisfy the commutation relation $[\hat{a}_\lambda, \hat{a}_{\lambda'}] = 0, [\hat{a}_\lambda^\dagger, \hat{a}_{\lambda'}^\dagger] = 0, [\hat{a}_\lambda, \hat{a}_{\lambda'}^\dagger] = \delta_{\lambda,\lambda'}$. Note that for photons in a Raman process, their wave vectors are small compared to the reciprocal lattice vector (or the lattice constant is much smaller than optical wavelength). Below, we consider spontaneous Raman scattering, which assumes that there are initially no photons in the state of scattered photon λ'.

The non-interacting part of $\hat{H}$ is denoted as $\hat{H}_0$. I use the interaction picture, an intermediate representation between the Schrödinger picture and the Heisenberg picture, so that when we do perturbation theory, we do not need to carry around a non-interacting operator $e^{-i\hat{H}_0 t}$.

Define

$$|\Psi(t)\rangle_I = e^{i\hat{H}_0 t/\hbar}|\Psi(t)\rangle_S \tag{2.19}$$

where the subscript I represents a state in the interaction picture and S represents a state in the Schrödinger picture. $|\Psi\rangle$ specifies the phonon and electron states as well as the photon states. We then have

$$i\hbar\frac{\partial|\Psi(t)\rangle_I}{\partial t} = \hat{\Delta}_I(t)|\Psi(t)\rangle_I \tag{2.20}$$

where $\hat{\Delta}_I(t) = e^{i\hat{H}_0 t}\hat{\Delta}_S e^{-i\hat{H}_0 t}$. The evolution of $|\Psi(t)\rangle_I$ can be described by a unitary operator defined as below

$$|\Psi(t)\rangle_I = \hat{U}_I(t)|\Psi_I(0)\rangle \tag{2.21}$$

By the way, $|\Psi(0)\rangle_I = |\Psi(0)\rangle_S$. Based on this definition, we then have

$$i\hbar\frac{\partial\hat{U}_I(t)}{\partial t} = \hat{\Delta}_I(t)\hat{U}_I(t) \tag{2.22}$$

and if we do the integral,

$$\hat{U}_I(t) = 1 - \frac{i}{\hbar}\int_{t_0}^{t} dt'\hat{\Delta}_I(t')\hat{U}_I(t') \tag{2.23}$$

If we substitute the right-hand side $U_I(t')$ with the left-hand side and repeat, then to lower orders,

$$\hat{U}_I(t) = 1 + \left(\frac{-i}{\hbar}\right)\int_{t_0}^{t} dt_1 \hat{\Delta}_I(t_1) + \left(\frac{-i}{\hbar}\right)^2 \int_{t_0}^{t} dt_1 \hat{\Delta}_I(t_1) \int_{t_0}^{t_1} dt_2 \hat{\Delta}_I(t_2)$$
$$+ \left(\frac{-i}{\hbar}\right)^3 \int_{t_0}^{t} dt_1 \hat{\Delta}_I(t_1) \int_{t_0}^{t_1} dt_2 \hat{\Delta}_I(t_2) \int_{t_0}^{t_1} dt_3 \hat{\Delta}_I(t_3) \tag{2.24}$$

so that

$$|\Phi_i(t)\rangle_I = |\Phi_i(0)\rangle_I + \left(\frac{-i}{\hbar}\right)\sum_n |\Phi_n\rangle \int_{t_0}^{t} dt_1 \langle\Phi_n|\Delta_I(t_1)|\Phi_i\rangle_I$$
$$+ \left(\frac{-i}{\hbar}\right)^2 \sum_{mn} |\Phi_n\rangle \int_{t_0}^{t} dt_1 \int_{t_0}^{t_1} dt_2 \langle\Phi_n|\Delta_I(t_2)|\Phi_m\rangle_I \langle\Phi_m|\Delta_I(t_1)|\Phi_i\rangle_I$$
$$+ \left(\frac{-i}{\hbar}\right)^3 \sum_{mnl} |\Phi_n\rangle \int_{t_0}^{t} dt_1 \int_{t_0}^{t_1} dt_2 \int_{t_0}^{t_2} dt_3 \langle\Phi_n|\Delta_I(t_3)|\Phi_m\rangle_I$$
$$\times \langle\Phi_m|\Delta_I(t_2)|\Phi_l\rangle_I \langle\Phi_l|\Delta_I(t_1)|\Phi_i\rangle_I \tag{2.25}$$

The term that goes into Raman scattering cross section is the third-order perturbation term. We show this process in Fig. 2.1. It is easier to represent the intermediate electron-hole pair states such as the creation of $\mu\mathbf{k}$ electron state and simultaneous annihilation of $\nu\mathbf{k}'$ electron state with just α (and β, since we need to invoke two intermediate states). The initial electronic state can be denoted as 0. We also ignore the representation of the wave vector such as $\mathbf{q}$ and $\mathbf{k}$. The integration over time taking $t \to \infty$ will eventually give a scattering rate that is

$$\frac{2\pi}{\hbar^6}\epsilon_{\lambda'}\epsilon_{\lambda}\left\{\frac{\Pi^{\lambda}_{0\beta}\Pi^{\lambda'}_{\beta\alpha}\Xi_{\alpha,0}}{(-\omega_\lambda+\Omega+\omega_\beta)(\Omega+\omega_\alpha)} + \frac{\Pi^{\lambda}_{0\beta}\Xi_{\beta,\alpha}\Pi^{\lambda'}_{\alpha 0}}{(-\omega_\lambda+\Omega+\omega_\beta)(-\omega_\lambda+\omega_\alpha)}\right.$$
$$+ \frac{\Xi_{0,\beta}\Pi^{\lambda}_{\beta\alpha}\Pi^{\lambda'}_{\alpha 0}}{(-\omega_\lambda+\omega_{\lambda'}+\omega_\beta)(-\omega_\lambda+\omega_\alpha)}$$
$$\left. + \text{ terms that swap photon fields}\right\}\delta(\omega_{\lambda'}-\omega_\lambda-\Omega) \tag{2.26}$$

Terms that swap photon fields means changing λ and ω_λ to λ' and $-\omega_{\lambda'}$. I did not discuss the near-resonance cases, which means that the denominators in Eq. 2.26 can be zero. As discussed previously, the electron-phonon coupling $\Xi_{\alpha\beta}$ includes both long-range (Fröhlich or piezoelectric coupling) and short-range (deformation potential) interaction.

Usually, the Raman process where an incident photon emits a quanta of phonon energy and scatters is called Stokes scattering, and the process that absorbs a phonon quanta is called anti-Stokes scattering.

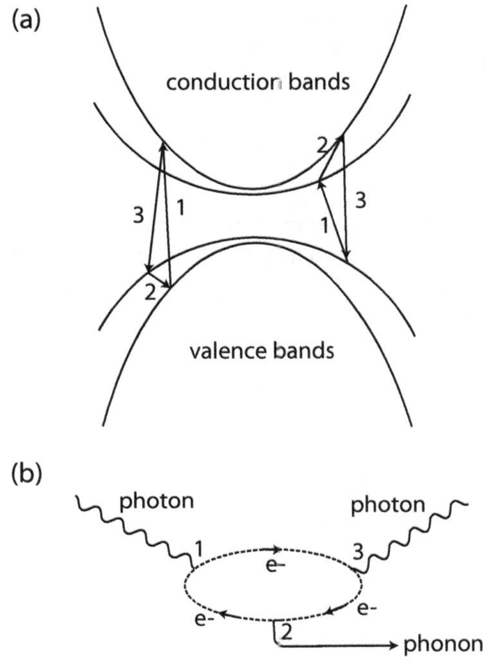

Fig. 2.1 (a) Electronic transitions and phonon emission (absorption) of a third-order light scattering process. The intraband (the process marked as 2) transition is due to phonons. Processes 1 and 3 are interband electronic transitions. The three processes can happen in any order, and there are six combinations. The figure is reproduced from ref. [13]. (b) The three-order process can also be represented with a Feynman diagram, where the electron-photon interaction is $\Pi^{\lambda\lambda'}_{\mu\mathbf{k}\nu\mathbf{k}'}$ (vertices 1 and 3) and the electron-phonon interaction is $\Xi^{\mu\nu}_{\mathbf{k}+\mathbf{q},\mathbf{k}}$ (vertex 2). $\Pi^{\lambda\lambda'}_{\mu\mathbf{k}\nu\mathbf{k}'}$ and $\Xi^{\mu\nu}_{\mathbf{k}+\mathbf{q},\mathbf{k}}$ are defined in the main text

2.3 Pump: Excitations of Phonons

In this section, we will discuss coherent excitations of phonons through the Raman process. I present below the microscopic theory of the stimulated Raman primarily following refs. [14, 15]. The incident light is not monochromated, but contains a spectrum, and there are abundant photons in the scattered Raman photon states even when $t \to -\infty$. It is thus easier for us to represent the light as a classical field that dresses the electronic states and produces time dependence in the term

$$\left\langle \sum_{\mu\nu,\mathbf{kq}} \Xi^{\mu\nu}_{\mathbf{k}+\mathbf{q},\mathbf{k}} c^{\dagger}_{\mu,\mathbf{k}+\mathbf{q}} c_{\nu,\mathbf{k}} \right\rangle, \tag{2.27}$$

which then provides the driving force of the phonon, which can also be considered classical when it is coherently driven. Only the electrons are treated as quantum mechanical: the dipole interaction between electrons and the light field is a perturbation to the initial electronic states. We want to obtain, eventually, an effective Hamiltonian where the light field couples directly to phonons, and the intermediating electrons are implicit. For off-resonant excitation, Eq. 2.27 is an average over the light-dressed electronic states. If the excitation is on-resonance (or above the bandgap), that is, real carriers are created with a lifetime much longer than the pulse duration, then one should think of Eq. 2.27 as averaging over the generated electron-hole pairs. Below, I will derive the phonon excitation based on the off-resonant excitation.

The interaction Hamiltonian in the interaction picture is

$$\hat{\Delta}_I(t) = |\alpha\rangle\langle\beta| \int d\omega e^{-i\omega_{\beta\alpha}t}\left[\Pi_{\beta\alpha}A(\omega)e^{-i\omega t} + \Pi^*_{\beta\alpha}A^*(\omega)e^{i\omega t}\right] \tag{2.28}$$

$A(\omega)$ is the amplitude of the classical light field. We have fixed the polarization of the photon field and absorbed it into $\Pi_{\beta\alpha}(k)$. Factors like $e^{-i\omega_{\beta\alpha}t}$ show up because we are using the interaction picture. We use $|\Phi\rangle_I$ to represent the electronic state in the interaction picture.

$$\begin{aligned}|\Phi(t)\rangle_I = |\Phi(0)\rangle_I &+ \left(\frac{-i}{\hbar}\right)\sum_\beta |\beta\rangle \int d\omega_1 \Bigg[A(\omega_1)\Pi_{0\beta}\int_{t_0}^{t} dt_1 e^{-i(\omega_{0\beta}+\omega_1)t_1} \\ &+ A^*(\omega_1)\Pi^*_{0\beta}\int_{t_0}^{t} dt_1 d^{-i(\omega_{0\beta}-\omega_1)t_1}\Bigg] + \left(\frac{-i}{\hbar}\right)^2 \sum_{\alpha\beta}|\beta\rangle \iint d\omega_1 d\omega_2 \\ &\Bigg[A(\omega_1)A^*(\omega_2)\Pi_{0\alpha}\Pi^*_{\alpha\beta}\int_{t_0}^{t} dt_1 e^{-i(\omega_{0\alpha}+\omega_1)t_1}\int_{t_0}^{t_1} dt_2 e^{-i(\omega_{\alpha\beta}-\omega_2)t_2} \\ &+ A^*(\omega_1)A(\omega_2)\Pi^*_{0\alpha}\Pi_{\alpha\beta}\int_{t_0}^{t} dt_1 e^{-i(\omega_{0\alpha}-\omega_1)t_1}\int_{t_0}^{t_1} dt_2 e^{-i(\omega_{\alpha\beta}+\omega_2)t_2}\Bigg]\end{aligned} \tag{2.29}$$

where $\omega_{\alpha\beta} = \omega_\beta - \omega_\alpha$. Taking the limit that $t_0 \to -\infty$ and then integrating out the time variables and then transforming back into the Schrödinger picture to remove the fast-rotating phase factors of electronic states, it will be

$$\begin{aligned}|\Phi(t)\rangle_S = |\Phi(0)\rangle_S &+ \left(\frac{1}{\hbar}\right)\sum_\beta |\beta\rangle \int d\omega_1 \Bigg[A(\omega_1)\Pi_{0\beta}\frac{e^{-i\omega_1 t}}{(\omega_{0\beta}+\omega_1)} \\ &+ A^*(\omega_1)\Pi^*_{0\beta}\frac{e^{i\omega_1 t}}{(\omega_{0\beta}-\omega_1)}\Bigg] + \left(\frac{1}{\hbar}\right)^2 \sum_{\alpha\beta}|\beta\rangle \iint d\omega_1 d\omega_2 \\ &\Bigg[A(\omega_1)A^*(\omega_2)\Pi_{0\alpha}\Pi^*_{\alpha\beta}\frac{e^{-i\omega_1 t}}{(\omega_{0\alpha}+\omega_1)}\frac{e^{i\omega_2 t}}{(\omega_{\alpha\beta}-\omega_2)} \\ &+ A^*(\omega_1)A(\omega_2)\Pi^*_{0\alpha}\Pi_{\alpha\beta}\frac{e^{i\omega_1 t}}{(\omega_{0\alpha}-\omega_1)}\frac{e^{-i\omega_2 t}}{(\omega_{\alpha\beta}+\omega_2)}\Bigg]\end{aligned} \tag{2.30}$$

Eventually, in $\langle\Phi(t)|\sum_{\mu\nu,\mathbf{kq}}\Xi^{\mu\nu}_{\mathbf{k+q},\mathbf{k}}c^\dagger_{\mu,\mathbf{k+q}}c_{\nu,k}|\Phi(t)\rangle_S$, the (A^0) terms must vanish because the excited eigenstates are orthogonal to the ground eigenstate. (A^1) terms will average out, because optical frequency is much higher than the phonon frequencies $\omega_{\mathbf{q}}$, and in the low-energy effective Hamiltonian, we do not care about them. We are only left with (A^2) terms to the lowest order:

$$
\begin{aligned}
&\langle\Phi(t)|\sum_{\mu\nu,\mathbf{kq}}\Xi^{\mu\nu}_{\mathbf{k+q},\mathbf{k}}c^{\dagger}_{\mu,\mathbf{k+q}}c_{\nu,\mathbf{k}}|\Phi(t)\rangle_S\\
&=\frac{1}{\hbar^2}\sum_{\alpha,\beta}\iint d\omega_1 d\omega_2 A(\omega_1)A^*(\omega_2)(\epsilon_{1\lambda}\cdot\epsilon^*_{2\lambda'})e^{-i(\omega_1-\omega_2)t}\\
&\Big\{\langle 0|\sum_{\mu\nu,\mathbf{kq}}\Xi^{\mu\nu}_{\mathbf{k+q},\mathbf{k}}c^{\dagger}_{\mu,\mathbf{k+q}}c_{\nu,\mathbf{k}}|\beta\rangle_S\Pi^{\lambda}_{\beta\alpha}\Pi^{\lambda'}_{\alpha 0}\frac{1}{(\omega_{0\beta}+\omega_1)(\omega_{\alpha\beta}-\omega_2)}\\
&+\Pi^{\lambda}_{0\beta}\Pi^{\lambda'}_{\beta\alpha}\langle\alpha|\sum_{\mu\nu,\mathbf{kq}}\Xi^{\mu\nu}_{\mathbf{k+q},\mathbf{k}}c^{\dagger}_{\mu,\mathbf{k+q}}c_{\nu,\mathbf{k}}|0\rangle_S\frac{1}{(\omega_{0\beta}-\omega_1)(\omega_{\beta\alpha}+\omega_2)}\\
&+\Pi^{\lambda}_{0\beta}\langle\beta|\sum_{\mu\nu,\mathbf{kq}}\Xi^{\mu\nu}_{\mathbf{k+q},\mathbf{k}}c^{\dagger}_{\mu,\mathbf{k+q}}c_{\nu,\mathbf{k}}|\alpha\rangle_S\Pi^{\lambda'}_{\alpha 0}\frac{1}{(\omega_{0\alpha}-\omega_1))(\omega_{0\beta}-\omega_2)}\\
&+\text{terms that swap photon fields}\Big\}
\end{aligned}
\tag{2.31}
$$

Swapping fields means swapping $(\omega_1, A(\omega_1))$ with $(-\omega_2, A^*(\omega_2))$. Here, the first two terms in the $\{\cdot\}$ come from multiplying (A^2) and (A^0) terms, while the third term comes from multiplying two (A^1) terms. The expression in $\{\cdot\}$ of Eq. 2.31, when using $\Xi_{\beta,\alpha}$ to substitute $\langle\beta|\sum_{\mu\nu,\mathbf{kq}}\Xi^{\mu\nu}_{\mathbf{k+q},\mathbf{k}}c^{\dagger}_{\mu,\mathbf{k+q}}c_{\nu,\mathbf{k}}|\alpha\rangle_S$, can be related to the expression in $\{\cdot\}$ of Eq. 2.26, upon changing variables in the integral. See ref. [14]. The effective Hamiltonian for the phonon takes the form

$$
\hat{H}_{\text{eff}}=\sum_{\mathbf{q}}\hbar\omega_{\mathbf{q}}b^{\dagger}_{\mathbf{q}}b_{\mathbf{q}}-[F_{\mathbf{q}}(t)b_{\mathbf{q}}+c.c.],
\tag{2.32}
$$

where we use $F_{\mathbf{q}}(t)$ to represent the time-dependent function of the quantity $\langle\sum_{\mu\nu,\mathbf{kq}}\Xi^{\mu\nu}_{k+q,k}c^{\dagger}_{\mu,k+q}c_{\nu,k}\rangle_S$ in Eq. 2.31 which is also the quantity of Eq. 2.27. From the way $F(t)$ couples to $b_{\mathbf{q}}$ in the Hamiltonian, it is the driving force for the phonon. Importantly, Eq. 2.31 describes the stimulated Raman scattering process where the driving force for phonons is proportional to the field amplitude squared or the intensity $I(t)$ [15]. When the excitation is off-resonance and achieved with ultrashort pulses (approximated as $I(t)\approx\delta(t)$), the phonon driving mechanism is referred to as impulsive stimulated Raman scattering (ISRS) [16]. For above-gap excitation, the phonon driving mechanism is termed displacive excitation of coherent phonons (DECP). In DECP, the driving force is also expected to be proportional to the fluence or pulse energy of the pump laser, as the density of electron-hole pairs generated is proportional to the energy deposited into the sample before saturation. For a better understanding of the relaxation processes involved in the energy and momentum of photoexcited carriers, as well as the relaxation of the phonon driving force, I recommend referring to the following sources: [17, 18].

DECP and ISRS have different initial phases of the coherent phonon. In the stimulated Raman process, the light field generates a force that drives the phonon, which is a harmonic oscillator. We ignore the **q** subscript of phonon mode and have the equation of motion

$$\frac{d^2u}{d^2t} + \Omega^2 u = F(t) \tag{2.33}$$

where u is the phonon coordinate. Using the Green function to solve the equation, one gets

$$u(t) = \int_{-\infty}^{t} \frac{\sin[\Omega(t-\tau)]}{\Omega} F(\tau)d\tau. \tag{2.34}$$

When the excitation is off-resonance, the force acting on the system is impulsive compared to the timescales of phonon dynamics. In this case, the force is represented by a Dirac delta function, denoted as $F(t) = \delta(t)$, and the resulting motion of the system is given by $u(t) = \frac{\sin(\Omega t)}{\Omega}$, where Ω is the characteristic frequency of the phonon mode.

On the other hand, when the excitation is above resonance, the force experienced by the system is determined by the electron distribution in the conduction band and the holes in the valence band. This force can persist for a longer duration compared to the timescales of phonon dynamics. In this case, the driving force can be treated as a step function, and the solution to the equation of motion (Eq. 2.34) is given by $u(t) = \frac{1-\cos(\Omega t)}{\Omega^2}$.

In situations where the driving force is neither impulsive nor step function-like, the phonon displacements follow a sinusoidal function with a generic phase value, which may not necessarily be integer multiples of $\frac{\pi}{2}$.

Equation 2.32 describes the first-order effective coupling between light and phonons. However, the second-order coupling between photons and lattice vibrations can lead to an impulsive change in the phonon frequency, resulting in squeezed phonon states [19, 20]. Squeezed phonons differ from coherent phonons excited through DECP or ISRS in that they possess a constant mean value. In terms of physical interpretation, the transient modification of interatomic interactions leads to a sudden change in the dynamical matrix and, consequently, the phonon frequencies and normal modes in reciprocal space (**q** space).

The derivation of squeezed phonon generation presented below partially draws from the supplementary material of reference [21].

The atomic displacement is as follows:

$$\mathbf{u}_s(\mathbf{R}_m) = \frac{1}{\sqrt{NM_s}} \sum_{\mathbf{q},j} \sqrt{\frac{\hbar}{2\omega_{\mathbf{q},j}}} \boldsymbol{\epsilon}^s_{\mathbf{q},j} e^{i\mathbf{q}\cdot(\mathbf{R}_m+\boldsymbol{\tau}_s)} (b_{\mathbf{q}} + b^\dagger_{-\mathbf{q}}) \tag{2.35}$$

where the $\boldsymbol{\epsilon}^s_{\mathbf{q},j}$ is the eigenvector of the force matrix. After the photoexcitation, only the force matrix changed so that $\boldsymbol{\epsilon}_{\mathbf{q},j}$ changes but the atomic displacement $\mathbf{u}_s$ and the momentum $\boldsymbol{p}_s$ remain their pre-photoexcitation value. Making everything explicitly time-dependent, we have the time-dependent atomic displacement and momentum:

$$\mathbf{u}_s(\mathbf{R}_m, t) = \frac{1}{\sqrt{NM_s}} \sum_{\mathbf{q},j} \sqrt{\frac{\hbar}{2\omega_{\mathbf{q},j}}} \boldsymbol{\epsilon}^s_{\mathbf{q},j} e^{i\mathbf{q}\cdot(\mathbf{R}_m+\boldsymbol{\tau}_s)} (b_{\mathbf{q}j}(t) + b^{\dagger}_{-\mathbf{q}j}(t)) \tag{2.36}$$

and at the same time

$$\boldsymbol{p}_s(\mathbf{R}_m, t) = -i\sqrt{\frac{M_s}{N}} \sum_{\mathbf{q},j} \sqrt{\frac{\hbar\omega_{\mathbf{q},j}}{2}} \boldsymbol{\epsilon}^s_{\mathbf{q},j} e^{i\mathbf{q}\cdot(\mathbf{R}_m+\boldsymbol{\tau}_s)} (b_{\mathbf{q}j}(t) - b^{\dagger}_{-\mathbf{q}j}(t)) \tag{2.37}$$

We make the assumption that the new set of creation/annihilation operators is linear combinations of pre-excitation creation/annihilation operators $b_{\mathbf{q}j'}(t_{0-})$ and $b_{\mathbf{q}j'}(t_{0-})$:

$$b_{\mathbf{q}j}(t) = \sum_{j'} U_{\mathbf{q}jj'}(t) b_{\mathbf{q}j'}(t_{0-}) + V_{-\mathbf{q}jj'}(t) b^{\dagger}_{-\mathbf{q}j'}(t_{0-}) \tag{2.38}$$

t_0 is the time when the force matrix suddenly changes. We should have the relations $V_{-\mathbf{q}jj'} = V^*_{\mathbf{q}jj'}$ and $U_{-\mathbf{q}jj'} = U^*_{\mathbf{q}jj'}$.

The following part is a derivation of explicit expression for $U_{\mathbf{q}jj'}$ and $V_{-\mathbf{q}jj'}$. The time evolution of the annihilation operator can be obtained from $U_{\mathbf{q}jj'}(t) = U_{\mathbf{q}jj'}(t_{0-})e^{-i\omega_{\mathbf{q}j}(t-t_0)}$ and $V_{-\mathbf{q}jj'}(t) = V_{-\mathbf{q}jj'}(t_{0-})e^{-i\omega_{\mathbf{q}j}(t-t_0)}$, and for simplicity in the following, I will just write $U_{\mathbf{q}jj'}(t_{0-})$ as $U_{\mathbf{q}jj'}$, same for $V_{-\mathbf{q}jj'}$. Similarly, I replace $b_{\mathbf{q}j}(t_{0-})$ with $b_{\mathbf{q}j}$. From the continuity of $\boldsymbol{u}_s(\mathbf{R}_m, t)$ and $\boldsymbol{p}_s(\mathbf{R}_m, t)$, we have

$$\frac{1}{\sqrt{\omega^{(0)}_{\mathbf{q}j}}} \boldsymbol{\epsilon}^{s(0)}_{\mathbf{q}j} (b_{\mathbf{q}j} + b^{\dagger}_{\mathbf{q}j}) - \frac{1}{\sqrt{\omega^{(1)}_{\mathbf{q}j}}} \boldsymbol{\epsilon}^{s(1)}_{\mathbf{q}j} \sum_{j'} \left[(U_{\mathbf{q}jj'} + V_{-\mathbf{q}jj'})(b_{\mathbf{q}j'} + b^{\dagger}_{\mathbf{q}j'}) \right] = 0 \tag{2.39}$$

$$\sqrt{\omega^{(0)}_{\mathbf{q}j}} \boldsymbol{\epsilon}^{s(0)}_{\mathbf{q}j} (b_{\mathbf{q}j} - b^{\dagger}_{\mathbf{q}j}) - \sqrt{\omega^{(1)}_{\mathbf{q}j}} \boldsymbol{\epsilon}^{s(1)}_{\mathbf{q}j} \sum_{j'} \left[(U_{\mathbf{q}jj'} - V_{-\mathbf{q}jj'})(b_{\mathbf{q}j'} - b^{\dagger}_{\mathbf{q}j'}) \right] = 0 \tag{2.40}$$

I use $\boldsymbol{\epsilon}^{s(1)}_{\mathbf{q}j}$ to represent eigenvectors diagonalized from the photoexcited force matrix and $\boldsymbol{\epsilon}^{s(0)}_{\mathbf{q}j}$ to represent eigenvectors diagonalized from the pre-excitation force matrix. Multiplying both sides of Eqs. 2.39 and 2.40 with $\boldsymbol{\epsilon}^{s(1)*}_{\mathbf{q}j}$, summing over s, applying the orthogonality relation $\sum_s (\boldsymbol{\epsilon}^s_{\mathbf{q},j})^* \cdot \boldsymbol{\epsilon}^s_{\mathbf{q},j'} = \delta_{j,j'}$, and then doing the operation $[\cdot, b^{\dagger}_{\mathbf{q}i}(t_0)]$ and $[\cdot, b_{\mathbf{q}i}(t_0)]$ separately to the equation, we get

$$U_{\mathbf{q}jj'} = \left[\sum_s \boldsymbol{\epsilon}^{s(0)}_{\mathbf{q}j'} \cdot \boldsymbol{\epsilon}^{s(1)*}_{\mathbf{q}j}\right] \frac{1}{2} \frac{\omega^{(0)}_{\mathbf{q}j'} + \omega^{(1)}_{\mathbf{q}j}}{\sqrt{\omega^{(0)}_{\mathbf{q}j'}\omega^{(1)}_{\mathbf{q}j}}} \tag{2.41}$$

which is only nonzero if the frequencies of phonons change after excitation, and

$$V_{-\mathbf{q}jj'} = \left[\sum_s \boldsymbol{\epsilon}^{s(0)}_{\mathbf{q}j'} \cdot \boldsymbol{\epsilon}^{s(1)*}_{\mathbf{q}j}\right] \frac{1}{2} \frac{-\omega^{(0)}_{\mathbf{q}j'} + \omega^{(1)}_{\mathbf{q}j}}{\sqrt{\omega^{(0)}_{\mathbf{q}j'}\omega^{(1)}_{\mathbf{q}j}}} \tag{2.42}$$

which is only nonzero if the eigenvectors of phonons change after photoexcitation. These time-dependent terms guarantee that the variance of a squeezed phonon state is not zero. In the end, Eq. 2.38 is a very different assumption from what we see in the effective Hamiltonian Eq. 2.32 where upon excitation, the oscillator does not change in frequency and its coupling with light is linear.

It is indeed worth noting that phonons can be coherently excited at nonzero wave vectors $\mathbf{q}$ in the presence of localized defects in a crystal. This phenomenon has been reported in the literature, with the first observation of such phonon generation reported in reference [22]. In Chap. 6 of the thesis, I will present findings on phonons generated through this mechanism. This demonstrates the ability to selectively excite specific phonon modes associated with localized defects in the crystal lattice.

2.4 Probe: X-Ray Scattering

The assumption that quantum electrodynamics dominates over the direct interaction between photons and nuclei is valid, given the significant difference in strength between electron-photon and photon-nuclei interactions. Building on this assumption, I will show that X-ray light effectively probes the charge density of electrons, which in turn reflects the dynamics of phonons.

It is worth noting that there are other types of scattering techniques, such as neutron scattering, which rely on different interaction mechanisms, specifically neutron-nuclei interactions. When it comes to the formalism of linear response theory and the fluctuation-dissipation theorem, the framework is quite similar to X-ray scattering despite the differences in the specific interaction mechanisms involved. A good reference of general inelastic scattering is Sinha et al. [23].

A General Formalism

Hard X-rays work differently than optical photons in terms of probing phonon dynamics. We still need to consider the light-electron interaction as in the previous section; however, for hard X-rays, we do not rely on the dipole approximation. Due to the large momentum of hard X-rays that match the scale of reciprocal lattice vectors, we probe the lattice dynamics making use of the significant phase change experienced by X-rays as they propagate across a unit cell of the lattice. Since we talk about lattice dynamics which happens at finite frequency, I will formalize X-ray scattering without assuming it is elastic. As is the case for optical probe, the properties of the X-ray probe pulse, such as its coherence and pulse duration, need to be taken into account in the scattering process. In the end, we will find that taking the limits of pulse shape and X-ray detector bandwidth, we get inelastic scattering either in the time domain or in the frequency domain—though experimentally these two types of experiments require very different instrumental capabilities. As mentioned, time-domain probing has the unique advantage in investigating the non-equilibrium states of matter as it matches the timescale of interesting non-equilibrium dynamics. Below, I largely follow the derivations in refs. [24–26].

As I will be discussing electron density in space later, the electrons will not be quantized in the spatial Fourier domain, which is the case in Eqs. 2.16 and 2.18. The light-electron interaction Hamiltonian with minimal coupling, in atomic units, is

$$\hat{H}_{int} = \alpha \int d^3x \hat{\psi}^\dagger(\mathbf{r})[\hat{A}(\mathbf{r}) \cdot \frac{\nabla}{i}]\hat{\psi}(\mathbf{r}) + \frac{\alpha^2}{2} \int d^3\mathbf{r} \hat{\psi}^\dagger(\mathbf{r}) \hat{A}^2(\mathbf{r}) \hat{\psi}(\mathbf{r}). \tag{2.43}$$

ψ is the spinor field operator for electrons and satisfies the anticommutation relations, $\{\hat{\psi}_\sigma(\mathbf{r}), \hat{\psi}_{\sigma'}(\mathbf{r}')\} = 0$, $\{\hat{\psi}^\dagger_\sigma(\mathbf{r}), \hat{\psi}^\dagger_{\sigma'}(\mathbf{r}')\} = 0$, $\{\hat{\psi}_\sigma(\mathbf{r}), \hat{\psi}^\dagger_{\sigma'}(\mathbf{r}')\} = \delta_{\sigma,\sigma'}\delta(\mathbf{r} - \mathbf{r}')$, where σ and σ' represent the spinor component. The light field is

$$\hat{A}(\mathbf{r}) = \sum_{\mathbf{k},\lambda} \sqrt{\frac{2\pi}{V\omega_{\mathbf{k}}\alpha^2}} \{\hat{a}_{\mathbf{k},\lambda}\epsilon_{\mathbf{k},\lambda} e^{i\mathbf{k}\cdot\mathbf{r}} + \hat{a}^\dagger_{\mathbf{k},\lambda}\epsilon^*_{\mathbf{k},\lambda} e^{-i\mathbf{k}\cdot\mathbf{r}}\} \tag{2.44}$$

V is the volume of the space, and α is the fine structure constant. The operators $\hat{a}_{\mathbf{k},\lambda}$ and $\hat{a}^\dagger_{\mathbf{k},\lambda}$ are the annihilation and creation operators of a photon in the mode $(\mathbf{k}, \lambda)$. $\mathbf{k}$ denotes the wave vector of the photon and λ the polarization.

A scattering event involves the initial state $|i\rangle$ and final state$|f\rangle$. The transition amplitude is

$$S = -2\pi i \delta(E_f - E_i) \times \left\{ \langle f|\hat{H}_{int}|i\rangle + \sum_m \frac{\langle f|\hat{H}_{int}|m\rangle\langle m|\hat{H}_{int}|i\rangle}{E_i - E_m + i\epsilon} + \ldots \right\} \tag{2.45}$$

Non-resonant X-ray scattering is due to the first term $\langle f|\hat{H}_{int}|i\rangle$ in Eq. (2.45). In the first-order perturbation, the non-resonance scattering is dominated by the second term in Eq. (2.43). This is because the first term in Eq (2.43) describes absorption or emission of photons, instead of scattering of photon from one state to another state. A more rigorous derivation will reveal this fact through the properties of creation and annihilation operator, which will not be expanded on here, corrections to scattering amplitude due to resonances come from the second term in Eq. (2.45). The wavelength window of X-rays I work with (9.5 KeV) is well above most resonances in the materials I study (except the K resonance that lies far above). For this reason, I neglect the resonant scattering term.

Probing with a non-resonant scattering process, of course, involves inputting a physical beam that is not a plane wave and has finite coherence and limited pulse duration as well as bandwidth. It also involves using a detector with limited detection bandwidth and energy resolution. The first issue can be addressed by separating out a term of the cross section that purely describes the light field properties, as we show below. The second issue can be addressed by introducing a projection operator over part of the Hilbert space

$$\hat{O}_{\mathbf{k}_f} = \sum_{j,\{n\}} |\Phi_j\{n\}\rangle\langle\Phi_j\{n\}| W_{\Delta\omega}(\omega_{\mathbf{k}_f}) \tag{2.46}$$

where Φ_j represents the electron energy states and $\{n\}$ represents the light field Fock state.

The differential scattering probability can be written as

$$\frac{dP}{d\Omega} = \frac{V\alpha^3}{(2\pi)^3} \int d\omega_{\mathbf{k}_f} \omega_{\mathbf{k}_f}^2 \operatorname{Tr}[\hat{O}_{\mathbf{k}_f} \hat{\rho}_f] \tag{2.47}$$

where $\hat{\rho}_f$ is the final state density matrix. Below, we will work in the interaction picture. It can be shown that [25]

$$\begin{aligned}
\frac{dP}{d\Omega} = &\sum_{\mathbf{k}_1,\lambda_1,\mathbf{k}_2,\lambda_2} \int_{-\infty}^{\infty}\int_{-\infty}^{\infty} dt_1 dt_2 \int_0^{\infty} d\omega_{\mathbf{k}_f} W_{\Delta\omega}(\omega_{\mathbf{k}_f}) \\
&\times \frac{\omega_{\mathbf{k}_f}\alpha^3}{2\pi V\sqrt{\omega_{\mathbf{k}_1}\omega_{\mathbf{k}_2}}} (\epsilon_{\mathbf{k}_1,\lambda_1}\cdot\epsilon^*_{\mathbf{k}_f,\lambda_f})(\epsilon_{\mathbf{k}_2,\lambda_2}\cdot\epsilon^*_{\mathbf{k}_f,\lambda_f}) \\
&\times \Bigg[\int d^3\mathbf{r} \int d^3\mathbf{r}' \langle\Phi_0, t_2|\hat{n}(\mathbf{r}')\hat{U}(t_2,t_1)\hat{n}(\mathbf{r})|\Phi_0, t_1\rangle \\
&\times e^{-i(\mathbf{k}_2-\mathbf{k}_f)\cdot\mathbf{r}'+i(\mathbf{k}_1-\mathbf{k}_f)\cdot\mathbf{r}-i\omega_{\mathbf{k}_1}t_1+i\omega_{\mathbf{k}_2}t_2-i\omega_{\mathbf{k}_f}(t_2-t_1)} \\
&\times \left(\sum_{\{m\}} \sum_{\{n\},\{\bar{n}\}} \hat{\rho}^i_{\{n\},\{\bar{n}\}} \langle\{\bar{n}\}|\hat{a}^\dagger_{\mathbf{k}_2,\lambda_2}|\{m\}\rangle \times \langle\{m\}|\hat{a}_{\mathbf{k}_1,\lambda_1}|\{n\}\rangle \right) \Bigg]
\end{aligned} \tag{2.48}$$

$\hat{\rho}^i$ is the initial state density matrix. $|\Phi_0, t_1\rangle$ represents the $|\Phi_0, t_0\rangle$ initial electronic state free evolved (assuming no interaction) to time t_1. $\hat{\psi}^\dagger(\mathbf{r})\hat{\psi}(\mathbf{r}) = \hat{n}(\mathbf{r})$ where $\hat{n}(\mathbf{r})$ is the electron density operator.

Equation 2.48 can be simplified as follows. First, there is a Thompson scattering (scattering from a single free electron) cross section

$$\frac{d\sigma_{th}}{d\Omega} = \alpha^2 \left(\sum_{\lambda_f=1}^{2} |\epsilon^*_{\mathbf{k}_i,\lambda_i} \cdot \epsilon_{\mathbf{k}_f,\lambda_f}|^2 \right). \tag{2.49}$$

Second, the last two rows of Eq. 2.48 only reflect the properties of the incident light field. We can lump everything related to the incident light field properties into a term $G^{(1)}(\mathbf{r}', \tau + \frac{\delta}{2}; \mathbf{r}, \tau - \frac{\delta}{2})$, where $\tau = \frac{t_1+t_2}{2}$ and $\delta = t_2 - t_1$. $G^{(1)}$ can be understood as the first-order correlation function of the X-ray probe beam. Then the non-resonant differential scattering probability for time-domain probing is

$$\begin{aligned}\frac{dP}{d\Omega} = &\frac{d\sigma_{th}}{d\Omega} \int_0^{\infty} d\omega_{\mathbf{k}_f} W_{\Delta\omega}(\omega_{\mathbf{k}_f}) \frac{\omega_{\mathbf{k}_f}}{\alpha(2\pi\omega_{\mathbf{k}_i})} \\ &\times \int_{-\infty}^{\infty} d\tau \int_{-\infty}^{\infty} d\delta e^{-i\omega_{\mathbf{k}_f}\delta} \int d^3\mathbf{r} \int d^3\mathbf{r}' \\ &\left\langle \Phi_0, \tau + \frac{\delta}{2} | \hat{n}(\mathbf{r}') e^{-i\hat{H}\delta} \hat{n}(\mathbf{r}) | \Phi_0, \tau - \frac{\delta}{2} \right\rangle e^{-i\mathbf{k}_f\cdot(\mathbf{r}-\mathbf{r}')} \\ &\times G^{(1)}(\mathbf{r}', \tau + \frac{\delta}{2}; \mathbf{r}, \tau - \frac{\delta}{2})\end{aligned} \tag{2.50}$$

Here the $\hat{H}$ is an electronic Hamiltonina which does not involve photons. If the X-ray probe beam is a coherent ensemble of Gaussian pulses, $G^{(1)}(\mathbf{r}', \tau + \frac{\delta}{2}; \mathbf{r}, \tau - \frac{\delta}{2}) = I(\tau)C(\delta)e^{i\omega_{\mathbf{k}_i}\delta}e^{i\mathbf{k}_i(\mathbf{r}-\mathbf{r}')}$, where $I(\tau)$ is the intensity of the probe X-ray pulse and $C(\delta) = \exp[-\frac{\delta^2}{\tau_l^2}]$ is a Gaussian function (τ_l can be understood as the probe pulse duration). We assumed that the incident field is assumed to be monochromatic enough and has a mean wave vector $\epsilon_{\mathbf{k}_i}$. If the short pulse has a much shorter time duration than the dynamics of electrons that we care about, we can approximate $C(\delta)$ as $F_0\delta(\tau - \tau_d)$ (F_0 is the fluence of the pulse). The short pulse launches coherent wave packets in the material. For the cases I study, these are phonon wave packets. The laser pulse must be shorter than all energy levels of the relevant wave packets E_i to launch a coherent wave packet. Therefore, if we denote the mean energy of the wave packet as $\langle \hat{H} \rangle$, we have $|E_i - \langle \hat{H} \rangle| \ll \frac{1}{\tau_l}$. We can then approximate that the unitary operator $e^{-i\hat{H}t}$ acting on any eigenstates $|E_i\rangle$ is equal to $e^{-i\langle \hat{H} \rangle t}$. Integrating over τ, the differential scattering probability is then

$$\begin{aligned}\frac{dP}{d\Omega} = &\frac{d\sigma_{th}}{d\Omega} F_0 \int_0^{\infty} d\omega_{\mathbf{k}_f} W_{\Delta\omega}(\omega_{\mathbf{k}_f}) \\ &\times \frac{\omega_{\mathbf{k}_f}}{\omega_{\mathbf{k}_i f}} \int_{-\infty}^{\infty} \frac{d\delta}{2\pi} C(\delta) e^{-i(\omega_{\mathbf{k}_f} - \omega_{\mathbf{k}_i})\delta} \int d^3\mathbf{r} \int d^3\mathbf{r}'\end{aligned}$$

$$\left\langle \Phi_0, \tau_d | \hat{n}(\mathbf{r}') e^{-i(\hat{H}-\langle \hat{H} \rangle)\delta} \hat{n}(\mathbf{r}) | \Phi_0, \tau_d \right\rangle e^{-i\mathbf{Q}\cdot(\mathbf{r}-\mathbf{r}')} \tag{2.51}$$

I have defined $\mathbf{Q} = \mathbf{k}_f - \mathbf{k}_i$. To simplify the expression, one can insert a complete basis $\sum_j |\Phi_j\rangle\langle\Phi_j|$, which allow us to convert the operator $\hat{H}$ into a quantum mechanical mean value and have it integrated out. Now if we have a detector that does not resolve energy, which means $W_{\Delta E}(\omega_{\mathbf{k}_f}) \equiv 1$, then eventually we end up with

$$\frac{dP(\mathbf{Q}, \tau_d)}{d\Omega} = \frac{d\sigma_{th}}{d\Omega} F_0 \int d^3\mathbf{r} \int d^3\mathbf{r}' \left\langle \Phi_0, \tau_d | \hat{n}(\mathbf{r}') \hat{n}(\mathbf{r}) | \Phi_0, \tau_d \right\rangle e^{-i\mathbf{Q}\cdot(\mathbf{r}-\mathbf{r}')}, \tag{2.52}$$

or equivalently, switching to the Heisenberg picture, we have

$$\frac{dP(\mathbf{Q}, \tau_d)}{d\Omega} = \frac{d\sigma_{th}}{d\Omega} F_0 \int d^3\mathbf{r} \int d^3\mathbf{r}' \left\langle \Phi_0 | \hat{n}(\mathbf{r}', \tau_d) \hat{n}(\mathbf{r}, \tau_d) | \Phi_0 \right\rangle e^{-i\mathbf{Q}\cdot(\mathbf{r}-\mathbf{r}')}. \tag{2.53}$$

The scattering in the non-energy-resolving detector and short pulse limit is not explicitly inelastic. It is the equal time density correlation that yields the scattering signal.

The other limit of the probing process, which is high detector energy resolution with monochromatic long pulses, will reveal the inelastic nature of the scattering. The steps leading to Eq. 2.48 include terms such as $\langle \{n'\} | \hat{a}_{\mathbf{k}_1,\lambda_1} \hat{a}^\dagger_{\mathbf{k}_f,\lambda_f} + \hat{a}_{\mathbf{k}_f,\lambda_f} \hat{a}^\dagger_{\mathbf{k}_1,\lambda_1} | \{n\} \rangle e^{i(E_{\{n'\}} - E_{\{n\}})t_1}$. When the exponent takes the $t_1 \to \infty$ limit, energy conservation law will be explicit as a delta function, meaning that to produce a real excitation in the electron density characterized by a finite frequency, a photon needs to lose energy. Alternatively, one can see this by letting $\tau_l \to \infty$ for $C(\delta) = \exp[-\frac{\delta^2}{\tau_l^2}]$ after inserting a complete basis of energy eigenstates $\sum_j |\Phi_j\rangle\langle\Phi_j|$ on the right-hand side of Eq. 2.51. The differential scattering probability will then be

$$\frac{dP(\mathbf{Q}, \omega)}{d\Omega} \propto \sum_j \delta(\omega + E_0 - E_j) \Big| \int d^3\mathbf{r} \langle \Phi_j | \hat{n}(\mathbf{r}) | \Phi_0 \rangle e^{-i\mathbf{Q}\cdot\mathbf{r}} \Big|^2 \tag{2.54}$$

which is how people do inelastic scattering in the frequency domain by using a highly monochromatic beam (typically in a thermal equilibrium setting), which necessarily means that the probe must have a long pulse duration. Apart from the two limits where one has either good time resolution or good energy resolution, there can be a region where one can probe with both sufficient time and frequency resolution.

In general, the excitations in materials can be probed either in the frequency domain when there is high energy resolution or in the time domain when there is enough bandwidth of the pump and probe pulses.

I list several assumptions that we make when probing phonon dynamics in the time domain with inelastic (hard) X-ray scattering:

(1) Phonon dynamics are much slower than most electronic dynamics (the period of phonon oscillations are much longer than that of electronic transitions), and the bandwidths of the incoming pulses are not a concern for us.
(2) Electron scattering is coherent, i.e., scattering from all the atoms coherently add up. There is no extra phase difference between atomic sites that comes from the electronic excitations.
(3) Core electrons always synchronously move with the atom (Born-Oppenheimer approximation), and it is these electrons that reflect the nuclei's motion.

If one wants to probe electronic transitions of much higher energy than lattice dynamics, in the time domain, one needs extremely short light pulses or extremely broad bandwidths, which are not readily available. Therefore, people who are (only) interested in high-energy electronic excitations might still prefer frequency-domain probe.

X-Ray Scattering in the Equilibrium

Suppose we freeze the time parameter τ_d in Eq. 2.53 as we want to understand the technique better before we look into the dynamics. We want to know at a given point τ_d in time what we expect to see given the electron configurations. Dropping the variable τ_d, we have

$$\frac{dP(\mathbf{Q})}{d\Omega} = \frac{d\sigma_{th}}{d\Omega} F_0 \Big| \int d^3\mathbf{r} \langle \Phi_0 | \hat{n}(\mathbf{r}) | \Phi_0 \rangle e^{-i\mathbf{Q}\cdot\mathbf{r}} \Big|^2 \tag{2.55}$$

We let $\langle n \rangle = \langle \Phi_0 | \hat{n}(\mathbf{r}) | \Phi_0 \rangle$ and just treat it as a classical quantity. Then there is

$$\frac{dP(\mathbf{Q})}{d\Omega} \propto \Big| \int d^3\mathbf{r} \langle n \rangle e^{-i\mathbf{Q}\cdot\mathbf{r}} \Big|^2 \tag{2.56}$$

We assume that we have a crystal with lattice sites $\mathbf{R}_m$. Each unit cell has atoms at different internal position $\boldsymbol{\tau}_s$ (note that $\boldsymbol{\tau}$ does not represent time here but the position. Each atom s has core electron charge density $n_s(\mathbf{r})$ around it. The total charge density will be $n(\mathbf{r}) = \sum_{m,s} n_s(\mathbf{r} - \mathbf{R}_m - \boldsymbol{\tau}_s)$. So then the scattering probability will be proportional to

$$\frac{dP(\mathbf{Q})}{d\Omega} \propto \Big| \int d^3\mathbf{r} \sum_{m,s} n_s(\mathbf{r} - \mathbf{R}_m - \boldsymbol{\tau}_s) e^{-i\mathbf{Q}\cdot\mathbf{r}} \Big|^2 \tag{2.57}$$

Making a transformation of integral variable $\boldsymbol{\xi} = \mathbf{r} - \mathbf{R}_m$,

$$\int d^3\mathbf{r} \sum_{m,s} n_s(\mathbf{r} - \mathbf{R}_m - \boldsymbol{\tau}_s) e^{-i\mathbf{Q}\cdot\mathbf{r}} = \sum_m e^{-i\mathbf{Q}\cdot\mathbf{R}_m} \sum_s \int d\boldsymbol{\xi} n_s(\boldsymbol{\xi} - \boldsymbol{\tau}_s) e^{-i\mathbf{Q}\cdot\boldsymbol{\xi}} \tag{2.58}$$

We can make

$$f_s(\mathbf{Q}) = \int d\boldsymbol{\xi} n_s(\boldsymbol{\xi}) e^{-i\mathbf{Q}\cdot\boldsymbol{\xi}} \tag{2.59}$$

We usually call $f_s(\mathbf{Q})$ a form factor for atom s. We also define the structure factor as below

$$S(\mathbf{Q}) = \sum_s f_s(\mathbf{Q}) e^{-i\mathbf{Q}\cdot\boldsymbol{\tau}_s} \tag{2.60}$$

so that

$$\frac{dP(\mathbf{Q})}{d\Omega} \propto \left|\sum_m S(\mathbf{Q}) e^{-i\mathbf{Q}\cdot\mathbf{R}_m}\right|^2 \tag{2.61}$$

One can think of $S(\mathbf{Q})$ as the scattered electric field by a single unit cell.

For perfect infinite ($N \to \infty$) crystals, we can proof that $\sum_m e^{-i\mathbf{Q}\cdot\mathbf{R}_m} = N\delta_{\mathbf{Q},\mathbf{G}}$ where $\mathbf{G}$ is the reciprocal lattice vector and N is the total number of unit cells in the crystal. This is simply done by enforcing $\sum_m e^{i\mathbf{Q}\cdot\mathbf{R}} = \sum_m e^{i\mathbf{Q}\cdot(\mathbf{R}+\mathbf{R}_0)}$ for arbitrary Bravais lattice $\mathbf{R}_0$. To the first order,

$$\frac{dP(\mathbf{Q})}{d\Omega} \propto N^2 \delta_{\mathbf{Q},\mathbf{G}} \left|S(\mathbf{Q})\right|^2 \tag{2.62}$$

In other words, the incident wave $\mathbf{k}_i$ gets scattered to $\mathbf{k}_f = \mathbf{k}_i + \mathbf{Q} = \mathbf{k}_i + \mathbf{G}$ and is intense ($\propto N^2$). Intensity does not scale as $\propto N^2$ for all other $\mathbf{Q} \neq \mathbf{G}$. But the crystals I study are far from perfect. In equilibrium, disorder, defects, and thermal excitations can all be causes of imperfection. Given these circumstances, the scattered intensity will not go extinct for $\mathbf{Q} \neq \mathbf{G}$.[2] But still, when $\mathbf{Q} = \mathbf{G}$, the scattering is still much brighter than $\mathbf{Q} \neq \mathbf{G}$. We call this certain X-ray scattering satisfying $\mathbf{Q} = \mathbf{G}$ geometry "diffraction."

I introduce the concept of the Ewald sphere below, which serves as a nice visual tool for Eq. 2.62. The rationale for building the Ewald sphere is that we can approximate the scattering as elastic, as explained in the previous subsection. In Fig. 2.2, the Ewald sphere (broken line) defines a sphere of scattered light wave vector $\mathbf{k}_f$ that satisfies $|\mathbf{k}_f| = |\mathbf{k}_i|$. One can imagine that if we rotate the crystal, the orange lattices (the reciprocal space lattices) will move relative to the Ewald sphere.

[2] The above is assuming kinematic diffraction. For high-quality crystals, we may need to consider electromagnetic waves exchanging between different modes $\mathbf{k}_i$ and $\mathbf{k}_f = \mathbf{k}_i + \mathbf{G}$. This is called dynamical diffraction [27].

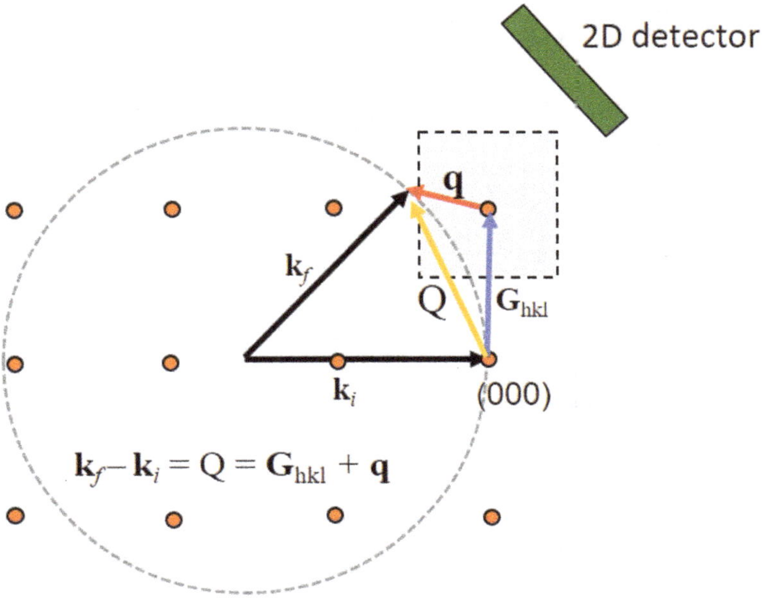

Fig. 2.2 Illustration of the Ewald sphere. The Ewald sphere (gray broken line) defines a sphere of scattered light wave vector $\mathbf{k}_f$ that satisfies $|\mathbf{k}_f| = |\mathbf{k}_i|$. Orange points are the reciprocal space Brillouin zone center. The Brillouin zone boundaries are shown in black broken lines. The blue vector represents $\mathbf{G}$ or more precisely $\mathbf{G}_{(hkl)}$ where (hkl) specifies the Miller indices. The squared shape gray box is the Brillouin zone of $\mathbf{G}_{(hkl)}$. The yellow vector represents $\mathbf{Q}$ the transfer momentum $\mathbf{k}_f - \mathbf{k}_i$. $\mathbf{q}$ in red arrow is the reduced wave vector in the given Brillouin zone. The 2D detector collects the scattered beam $\mathbf{k}_f$

This will not change the captured set of $\mathbf{k}_f$ by the 2D detector, but will change $\mathbf{G}_{(hkl)}$ and $\mathbf{q}$. (hkl) specifies the Miller indices of $\mathbf{G}$. The origin of the reciprocal space is (000). Only at specific crystal orientation will $\mathbf{Q}$ be overlapped with a $\mathbf{G}_{(hkl)}$ and $\mathbf{G}_{(hkl)}$ fall onto the Ewald sphere. That is when we get diffraction. We also call such crystal orientation as "hitting a Bragg condition."

In later chapters, as I present the experimental results, when I speak of diffraction, I mean we hit the Bragg condition of $\mathbf{G}_{(hkl)}$, and at the same time, the detector captures the scattered $\mathbf{k}_f$ transferred by $\mathbf{Q} = \mathbf{G}_{(hkl)}$. When I speak of diffuse scattering, I mean we don't hit Bragg condition, and I mainly care about dynamics at $\mathbf{Q} \neq \mathbf{G}_{(hkl)}$. Note that practically we don't do time-resolved diffraction and time-resolved diffuse scattering with the same sample geometry due to the fact that two types of signals are not simultaneously optimized. The diffuse scattering is orders of magnitude weaker than diffraction, and the detector has a limited dynamical range.

On top of a static perfect crystal, we now allow phonon excitations. We introduce the scattering intensity $I(\mathbf{Q})$ in the place of $\frac{dP}{d\Omega}$. The expression for the scattering intensity is

$$I(\mathbf{Q}) = I_0\langle|\sum_{m,s} f_s \exp[i\mathbf{Q}\cdot(\mathbf{R}_m + \boldsymbol{\tau}_s + \mathbf{u}_{m,s})]|^2\rangle \tag{2.63}$$

where $\mathbf{R}_m$ is the lattice site, the $\boldsymbol{\tau}_s$ is the position of atom in unit cell, and $\mathbf{u}_{m,s}$ is the displacement of the atom at $\mathbf{R}_m + \boldsymbol{\tau}_s$. Ignoring time dependence for now and treating phonons as classical fields,

$$\mathbf{u}_{m,s} = \frac{1}{2\sqrt{M_s}}\sum_{\mathbf{q}j} A_{\mathbf{q}j}\boldsymbol{\epsilon}^s_{\mathbf{q}j}\exp[i\mathbf{q}\cdot(\mathbf{R}_m + \boldsymbol{\tau}_s)] + A^*_{\mathbf{q}j}\boldsymbol{\epsilon}^{s*}_{\mathbf{q}j}\exp[-i\mathbf{q}\cdot(\mathbf{R}_m + \boldsymbol{\tau}_s)] \tag{2.64}$$

I_0 is related to the incident flux. M_s is the atomic mass. The bracket $\langle\cdot\rangle$ means thermal average in the thermal equilibrium in a classical picture or the expectation value in the quantum mechanical sense. As we are not in the low-temperature limit ($\bar{n} \approx \bar{n} + 1$; $\bar{n}$ is the expectation value of Boson occupation number), we do not consider the commutation relation when raising $\langle\cdot\rangle$ to the exponent.

$$\begin{aligned} I(\mathbf{Q}) = I_0\Big\langle \sum_{m,m',s,s'} &\big(f_s f^*_{s'}\exp[-i\mathbf{Q}\cdot\big(\mathbf{R}_{m,m'} + \boldsymbol{\tau}_{s,s'}\big)] \\ &\times \exp\Big\{-\frac{1}{2}\langle[\mathbf{Q}\cdot(\mathbf{u}_{m,s} - \mathbf{u}_{m',s'})]^2\rangle\Big\}\Big\rangle \end{aligned} \tag{2.65}$$

where I already used the approximation $\langle\exp(ix)\rangle = \exp(-\frac{1}{2}\langle x^2\rangle)$ if x is small with equal probability of being positive and negative or if it follows a Gaussian distribution [28].

$$\begin{aligned} I(\mathbf{Q}) = I_0 \sum_{m,m',s,s'} &f_s f_{s'} e^{-i\mathbf{Q}\cdot\left(\mathbf{R}_{m,m'}+\boldsymbol{\tau}_{s,s'}\right)} e^{-\frac{1}{2}\langle(\mathbf{Q}\cdot\mathbf{u}_{m,s})^2\rangle} \\ &\times e^{-\frac{1}{2}\langle(\mathbf{Q}\cdot\mathbf{u}_{m',s'})^2\rangle} e^{\langle(\mathbf{Q}\cdot\mathbf{u}_{ms})(\mathbf{Q}\cdot\mathbf{u}_{m's'})\rangle} \end{aligned} \tag{2.66}$$

The last exponential term can be Taylor expanded into

$$1 + \langle(\mathbf{Q}\cdot\mathbf{u}_{ms})(\mathbf{Q}\cdot\mathbf{u}_{m's'})\rangle + \frac{1}{2}\langle(\mathbf{Q}\cdot\mathbf{u}_{ms})(\mathbf{Q}\cdot\mathbf{u}_{m's'})\rangle^2 + \ldots \tag{2.67}$$

If $\mathbf{Q} = \mathbf{G}$, then the lowest order is

$$I(\mathbf{G}) = N^2 I_0\big|\sum_s f_s\exp(-i\mathbf{G}\cdot\boldsymbol{\tau}_s)\exp[-\frac{1}{2}\langle(\mathbf{G}\cdot\mathbf{u}_s)^2\rangle]\big|^2 \tag{2.68}$$

where I used $\sum_m e^{i\mathbf{Q}\cdot\mathbf{R}_m} = N\delta(\mathbf{G} - \mathbf{Q})$, which is proper for large crystal in the kinematic limit. N is the number of unit cells in the crystal. The terms such as

$\langle(\mathbf{Q}\cdot\mathbf{u}_{ms})^2\rangle$ should not depend on the unit cell site m, and we substitute $\mathbf{u}_{ms}$ with just $\mathbf{u}_s$.

If $\mathbf{Q}\neq\mathbf{G}$, the lowest-order diffuse scattering term is

$$I(\mathbf{Q}) = I_0 \sum_{m,m',s,s'} \tilde{f}_s \tilde{f}_{s'} \exp[-i\mathbf{Q}\cdot(\mathbf{R}_{m,m'}+\boldsymbol{\tau}_{s,s'})]\langle(\mathbf{Q}\cdot\mathbf{u}_{ms})(\mathbf{Q}\cdot\mathbf{u}_{m's'})\rangle \tag{2.69}$$

where

$$\tilde{f}_s = f_s e^{-\frac{1}{2}\langle(\mathbf{Q}\cdot\mathbf{u}_s)^2\rangle} \tag{2.70}$$

In the equilibrium, there is no fixed phase relation between different phonon modes j, j'; thus, the average term $\langle\cdot\rangle$ with $A_{\mathbf{q}j}A^*_{\mathbf{q}j'}$ will vanish if $j\neq j'$. We have, therefore,

$$I(\mathbf{Q}) = I_0 N \sum_{s,s',j} \frac{|A_{\mathbf{q}j}|^2}{\sqrt{M_s M_{s'}}} \tilde{f}_s \tilde{f}^*_{s'} \exp(-i\mathbf{G}\cdot\boldsymbol{\tau}_{s,s'})(\mathbf{Q}\cdot\epsilon^{s*}_{\mathbf{q}j})(\mathbf{Q}\cdot\epsilon^{s'}_{\mathbf{q}j})\big|_{\mathbf{q}=\mathbf{Q}-\mathbf{G}} \tag{2.71}$$

where we do not have terms like $(\mathbf{Q}\cdot\epsilon^{s*}_{\mathbf{q}j})(\mathbf{Q}\cdot\epsilon^{s'}_{\mathbf{q}j'})$, $(j\neq j')$. $|A_{\mathbf{q}j}|^2$ describes the thermal vibration amplitude and hence is related to the temperature. Based on the Bose-Einstein distribution, the thermal diffuse scattering intensity is

$$I(\mathbf{Q})_{thermal} = \frac{1}{2} I_0 \hbar N \sum_{j'} \frac{1}{\omega^{(0)}_{\mathbf{q}j'}} \coth\frac{\hbar\omega^{(0)}_{\mathbf{q}j'}}{2k_B T} \Big|\sum_s \frac{\tilde{f}_s}{\sqrt{M_s}} e^{-i\mathbf{G}\cdot\boldsymbol{\tau}_s}(\boldsymbol{\epsilon}^{s(0)}_{\mathbf{q}j}\cdot\mathbf{Q})\Big|^2 \tag{2.72}$$

The superscript (0) means in equilibrium or before excitation by the light pulse.

Time-Resolved Diffraction

Now I shall put back the τ_d in Eq. 2.53, as eventually, we do care about dynamics. In a pump-probe experiment, τ_d has a physical meaning, which is the time delay between the pump and probe pulses.

The atomic motion induced by the pump pulse is different from the thermal incoherent atomic motion. For the former case, we care about the phase of the phonon, and for the latter case, we do not. In the thermal equilibrium, one mostly cares about the diffraction intensity, while in a non-equilibrium setting, one cares more about the coherent modulation of diffraction intensity. We consider below the Bragg peak intensity changes incurred by atomic motion in real time.

We first consider zone center coherent phonons. The change of intensity will be more sensitive when atoms move along some coordinates than others. In fact, we can take the derivative of the structure factor to find out. If each unit cell contains only two atoms, each occupying $+\boldsymbol{\tau}$ and $-\boldsymbol{\tau}$, and the atomic motion takes each to new positions $\boldsymbol{\tau} + u\boldsymbol{\epsilon}$ and $-\boldsymbol{\tau} - u\boldsymbol{\epsilon}$, where ϵ is the phonon polarization and u is the coordinate of phonon mode ϵ, then

$$S(\mathbf{G}) = 2f_s(\mathbf{G}) \cos\left[\mathbf{G} \cdot (\boldsymbol{\tau} + u\boldsymbol{\epsilon})\right] \tag{2.73}$$

because every unit cell behaves exactly the same wave. Taking the derivative of the differential scattering probability with respect to u, one finds

$$\frac{\partial}{\partial u}\frac{dP}{d\Omega} \propto \frac{\partial S}{\partial u}S^* + \frac{\partial S*}{\partial u}S = -4N^2|f_s(\mathbf{G})|^2 \sin\left[2\mathbf{G} \cdot (\boldsymbol{\tau} + u\boldsymbol{\epsilon})\right](\mathbf{G} \cdot \boldsymbol{\epsilon}) \tag{2.74}$$

Although in a generic crystal, the atomic positions and atomic motion coordinates will be much more complicated, Eq. 2.74 is a simple demonstration that for a Bragg peak to be a sensitive probe for a certain vibrational mode $\boldsymbol{\epsilon}$, $\mathbf{G}$ is desired to have a large projection along the mode polarization due to the factor $(\mathbf{G} \cdot \boldsymbol{\epsilon})$. Below, I take bismuth as an example.

Bismuth has a rhombohedral unit cell containing two atoms. See Fig. 2.3. The polarization of the A_{1g} mode of bismuth is as shown with black arrows and fully respects the symmetry of the crystal. The simulations of normalized diffraction intensity as the function of phonon coordinate u are shown for $\mathbf{G}$ = (111), (002), (121), (1−10). Peaks that have large $\mathbf{G}$ projection on $\boldsymbol{\epsilon}$ have a higher sensitivity to the same phonon modulation $u\boldsymbol{\epsilon}$. If the mode vector $\boldsymbol{\epsilon}$ is known, one can predict which peaks are sensitive for probing the atomic motion (this step happens before

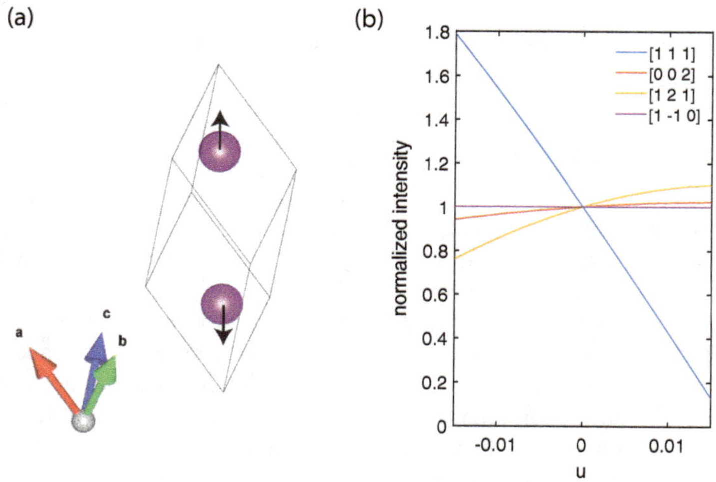

Fig. 2.3 (a) Bismuth structure and A_g modes. Black arrows represent A_g mode eigenvector $\boldsymbol{\epsilon}$. (b) Bismuth diffraction normalized intensity change with A_g coordinate u

the experiment), or by measuring the intensity modulation, one can calculate the atomic motion amplitude u (this step happens after the experiment, and we will give such an example in Chaps. 5 and 7). For crystals with more atoms per unit cell and unknown modes of possibly lower symmetry, one can reconstruct the mode vector ϵ by probing the intensity modulation u incurred by the mode on multiple different peaks $\mathbf{G}$. We will give an example for such procedure in Chap. 7.

Time-resolved diffraction can also probe squeezed phonon state because the form factor in Eq. 2.70 can be time-dependent:

$$\left\langle (\mathbf{G}\cdot\mathbf{u}_s)^2\right\rangle = \frac{\hbar}{NM_s}\sum_{\mathbf{q}j'}\langle\bar{n}\rangle_{\mathbf{q}j'}\left|\sum_j\sqrt{\frac{1}{\omega^{(1)}_{\mathbf{q}j}}}(\boldsymbol{\epsilon}^{s(1)}_{\mathbf{q}j}\cdot\mathbf{G})(U_{\mathbf{q}jj'}(t)+V_{-\mathbf{q}jj'}(t))\right|^2 \tag{2.75}$$

(Recall the previous section for derivation of squeezed phonon states.) This has indeed been observed in bismuth [21].

Time-Resolved Diffuse Scattering

In a perfect crystal, off the Bragg peak, the scattering due to phonons cannot be purely elastic, as phonons have both finite frequency and wave vector. Therefore, off Bragg, one can only see the beating between the Stokes and anti-Stokes components, which happens at twice the phonon frequency.

As mentioned, the lowest-order diffuse scattering is

$$\begin{aligned} I(\mathbf{Q}) = I_0\hbar N &\sum_{s,s',j'}\langle\bar{n}\rangle_{\mathbf{q}j'}\tilde{f}_s\tilde{f}^*_{s'}e^{-i\mathbf{G}\cdot\boldsymbol{\tau}_{s,s'}} \\ &\times\left|\sum_j\sqrt{\frac{1}{\omega^{(1)}_{\mathbf{q}j}}}(\boldsymbol{\epsilon}^{s(1)}_{\mathbf{q}j}\cdot\mathbf{Q})(U_{\mathbf{q}jj'}(t)+V_{-\mathbf{q}jj'}(t))\right|^2\Big|_{\mathbf{q}=Q-G} \end{aligned} \tag{2.76}$$

and this will give rise to the overtone of the phonons (diffuse scattering intensity modulated at twice the phonon frequency), because from Eqs. 2.41 and 2.42, putting in the phase factors $e^{-i\omega_{\mathbf{q}j}(t-t_0)}$ for $U_{\mathbf{q}jj'}(t)$ and $V_{-\mathbf{q}jj'}$, and letting $t_0=0$:

$$U_{\mathbf{q}jj'}(t)+V_{-\mathbf{q}jj'}(t) = \left[\sum_s\boldsymbol{\epsilon}^{s(0)}_{\mathbf{q}j'}\cdot\boldsymbol{\epsilon}^{s(1)*}_{\mathbf{q}j}\right]\left[\frac{\omega^{(1)}_{\mathbf{q}j}\cos\omega^{(1)}_{\mathbf{q}j}t - i\omega^{(0)}_{\mathbf{q}j'}\sin\omega^{(1)}_{\mathbf{q}j}t}{\sqrt{\omega^{(1)}_{\mathbf{q}j}\omega^{(0)}_{\mathbf{q}j'}}}\right] \tag{2.77}$$

Below, I briefly discuss how a non-perfect crystal with defects can give rise to heterodyned signal (scattering intensity modulated at the phonon frequency) at $\mathbf{Q}\neq\mathbf{G}$. I formulate the scattering intensity as an interference between two fields. The scattered field from an (almost) perfect crystal is

$$E_c(\mathbf{Q}) = \sum_{m,s} f_s \exp\left[i\mathbf{Q}\cdot(\mathbf{R}_m + \boldsymbol{\tau}_s + \mathbf{u}_{m,s})\right] \tag{2.78}$$

The atomic motion ignoring their time dependence, for now, is

$$\mathbf{u}_{m,s} = \sum A_{\mathbf{q},j}\boldsymbol{\epsilon}^s_{\mathbf{q},j}[\exp(i\mathbf{q}\cdot\mathbf{r}_{m,s}) + \exp(-i\mathbf{q}\cdot\mathbf{r}_{m,s})] \tag{2.79}$$

where $\mathbf{r}_{m,s} = \mathbf{R}_m + \boldsymbol{\tau}_s$.We ignored the square root mass term in Eq. 2.64. Therefore,

$$E_c(\mathbf{Q}) = \sum_{m,s} f_s \exp[i\mathbf{Q}\cdot\mathbf{r}_{m,s}]\exp[\sum_j iA_j(\mathbf{Q}\cdot\boldsymbol{\epsilon}^s_{\mathbf{q}j})[\exp(i\mathbf{q}\cdot\mathbf{r}_{m,s})+\exp(-i\mathbf{q}\cdot\mathbf{r}_{m,s})])] \tag{2.80}$$

Then using Taylor expansion,

$$E_c(\mathbf{Q}) = \sum_{m,s} f_s \exp[i\mathbf{Q}\cdot\mathbf{r}_{m,s}]\left(1 + \sum_j iA_j(\mathbf{Q}\cdot\boldsymbol{\epsilon}^s_{\mathbf{q}j})[\exp(i\mathbf{q}\cdot\mathbf{r}_{m,s}) + \exp(-i\mathbf{q}\cdot\mathbf{r}_{m,s})]\right) \tag{2.81}$$

As we are only interested in $\mathbf{q} \neq 0$ and $\mathbf{Q} - \mathbf{q} = \mathbf{G}$,

$$E_c(\mathbf{Q}) = N\sum_{s,j} if_s A_j(\mathbf{Q}\cdot\boldsymbol{\epsilon}^s_{\mathbf{q}j})\exp(i\mathbf{G}\cdot\boldsymbol{\tau}_s) \tag{2.82}$$

If there is no other field, the scattering intensity will just be

$$I(\mathbf{Q}) = |E_c(\mathbf{Q})|^2 = N^2\sum_{s,s',j} f_s f^*_{s'} A_{\mathbf{q}j}A_{-\mathbf{q}j}(\mathbf{Q}\cdot\boldsymbol{\epsilon}^s_{\mathbf{q}j}(\mathbf{q}))\left(\mathbf{Q}\cdot\boldsymbol{\epsilon}^{s'}_{\mathbf{q}j}(-\mathbf{q})\right) \times \exp[i\mathbf{G}\cdot(\boldsymbol{\tau}_s - \boldsymbol{\tau}_{s'})] \tag{2.83}$$

It is a sum of pair contribution $\langle A_{\mathbf{q}j}A_{-\mathbf{q}j}\rangle$ ($\langle a_{\mathbf{q}j}a_{-\mathbf{q}j}\rangle$ and $\langle b^\dagger_{\mathbf{q}j}b^\dagger_{-\mathbf{q}j}\rangle$).

With defects in the crystal,

$$E(\mathbf{Q}) = E_c(\mathbf{Q}) + E_D(\mathbf{Q}) \tag{2.84}$$

E_D is the scattered field from the defect. The exact form of E_D requires knowledge of details of defect distribution. With a lack of such information, we need to make some assumptions. We assume that the defects are identical small islands sparsely and randomly scattered around in the crystal. On each island, there is an elastic strain field so that atoms within that island are displaced away from a perfect crystal, and we can decompose such lattice distortion within a single island using the perfect crystal's phonon eigenmodes. We assume substitutions of atoms below. If we think of defects as

$$E_D(\mathbf{Q}) = f_s^D + \sum_s^{m\in\text{island}} f_s e^{i\mathbf{Q}\cdot\mathbf{r}_{m,s}} \times \left(1 + \sum_j i A_j^D (\mathbf{Q}\cdot\boldsymbol{\epsilon}_{\mathbf{q}j}^s)[\exp(i\mathbf{q}\cdot\mathbf{r}_{m,s}) + \exp(-i\mathbf{q}\cdot\mathbf{r}_{m,s})]\right) \tag{2.85}$$

f_s^D is the form factor difference with atoms that are part of the perfect crystal. $m \in$ island, which refers to unit cell position, enumerates through all atomic sites in the island. We do not consider the distribution of the islands, which will give rise to an extra scattering structure on top of what happens inside the islands. We emphasize that since we are describing static distortions, A_j^D has no time dependence.

We point out that although we don't have $\sum_m \exp(i\mathbf{Q}\cdot\mathbf{R}_m) = N\delta(\mathbf{Q}-\mathbf{G})$ if the crystal is not perfect and infinite, I assume that $\sum_{m\in\text{island}} \exp(i\mathbf{Q}\cdot\mathbf{R}_m) = N_D\Delta(\mathbf{Q}-\mathbf{G})$, where N_D is the total number of unit cells we consider for the continuous distortion field. Δ is a function peaked at 0 with finite width, and its integration over the whole domain is unity. We have the heterodyning term in the total scattering intensity

$$I(\mathbf{Q}) = E_c(\mathbf{Q})E_D^*(\mathbf{Q}) + c.c \tag{2.86}$$

$E_c(\mathbf{Q})$ is the scattered field from the perfect crystal, and $E_D(\mathbf{Q})$ is the scattered field from the defect island. The continuous elastic field gives rise to

$$I(\mathbf{Q}) \approx N N_D \sum_{jj'ss'} f_s f_{s'}^* \langle A_{\mathbf{q}j} A_{-\mathbf{q}j'}^D \rangle (\mathbf{Q}\cdot\boldsymbol{\epsilon}_j^{\mathbf{q}s})(\mathbf{Q}\cdot\boldsymbol{\epsilon}_{j'}^{\mathbf{q}s'*}) + c.c \tag{2.87}$$

while the random atomic substitution gives rise to

$$I(\mathbf{Q}) \approx N \sum_{s,j} f_s f^{D*} \langle A_{\mathbf{q}j} \rangle (\mathbf{Q}\cdot\boldsymbol{\epsilon}_{\mathbf{q}j}^s) + c.c \tag{2.88}$$

Both $\langle A_{\mathbf{q}j} A_{-\mathbf{q}j'}^D \rangle$ and $\langle A_{\mathbf{q}j} \rangle$ contain time dependence from $A_{\mathbf{q}j}$. The defects and the elastic deformation provide fields with finite momentum but with zero energy. If $A_{\mathbf{q}j}$ is a classical oscillator coherently excited at $\mathbf{q} \neq \mathbf{0}$, we can probe it without overtones.

References

1. K.M. Rabe, M. Dawber, C. Lichtensteiger, C.H. Ahn, J.-M. Triscone, Modern physics of ferroelectrics: Essential background, in *Physics of Ferroelectrics* (Springer, Berlin, 2007), pp. 1–30
2. M. Born, K. Huang, M. Lax, Dynamical theory of crystal lattices. Amer. J. Phys. **23**(7), 474–474 (1955)
3. H. Bilz, G. Benedek, A. Bussmann-Holder, Theory of ferroelectricity: The polarizability model. Phys. Rev. B **35**(10), 4840–4849 (1987)
4. A. Bussmann-Holder, H. Bilz, R. Vogl, Electronic and dynamical properties of iv–vi compounds, in *Dynamical Properties of IV-VI Compounds* (Springer, Berlin, 1983), pp. 51–98
5. M.J.P. Musgrave, J.A. Pople, A general valence force field for diamond. Proc. Roy. Soc. London Ser. A. Math. Phys. Sci. **268**(1335), 474–484 (1962)
6. R.M. Martin, Dielectric screening model for lattice vibrations of diamond-structure crystals. Phys. Rev. **186**(3), 871 (1969)
7. R.M. Martin, Elastic properties of zns structure semiconductors. Phys. Rev. B **1**(10), 4005 (1970)
8. J. Sólyom, *Fundamentals of the Physics of Solids: Volume 3-Normal, Broken-Symmetry, and Correlated Systems*, vol. 3 (Springer Science & Business Media, Cham, 2010)
9. R.M. Martin, Piezoelectricity. Phys. Rev. B **5**, 1607–1613 (1972)
10. P. Vogl, Microscopic theory of electron-phonon interaction in insulators or semiconductors. Phys. Rev. B **13**, 694–704 (1976)
11. M. Cardona, Y.Y. Peter, *Fundamentals of Semiconductors*, vol. 619 (Springer, Berlin, 2005)
12. R. Loudon, The raman effect in crystals. Adv. Phys. **13**(52), 423–482 (1964)
13. R. Loudon, W. Hayes, *Scattering of Light by Crystals*. (Wiley, Hoboken, 1978)
14. T.E. Stevens, J. Kuhl, R. Merlin, Coherent phonon generation and the two stimulated raman tensors. Phys. Rev. B **65**, 144304 (2002)
15. R. Merlin, Generating coherent THz phonons with light pulses. Solid State Commun. **102**(2), 207–220 (1997). Highlights in Condensed Matter Physics and Materials Science
16. Y.-X. Yan, E.B. Gamble Jr., K.A. Nelson, Impulsive stimulated scattering: general importance in femtosecond laser pulse interactions with matter, and spectroscopic applications. J. Chem. Phys. **83**(11), 5391–5399 (1985)
17. J.J. Li, J. Chen, D.A. Reis, S. Fahy, R. Merlin, Optical probing of ultrafast electronic decay in BI and SB with slow phonons. Phys. Rev. Lett. **110**, 047401 (2013)
18. S.M. O'Mahony, F. Murphy-Armando, É.D. Murray, J.D. Querales-Flores, I. Savić, S. Fahy, Ultrafast relaxation of symmetry-breaking photo-induced atomic forces. Phys. Rev. Lett. **123**, 087401 (2019)
19. G.A. Garrett, A.G. Rojo, A.K. Sood, J.F. Whitaker, R. Merlin, Vacuum squeezing of solids: macroscopic quantum states driven by light pulses. Science **275**(5306), 1638–1640 (1997)
20. M. Trigo, M. Fuchs, J. Chen, M.P. Jiang, M. Cammarata, S. Fahy, D.M. Fritz, K. Gaffney, S. Ghimire, A. Higginbotham, S.L. Johnson, M.E. Kozina, J. Larsson, H. Lemke, A.M. Lindenberg, G. Ndabashimiye, F. Quirin, K. Sokolowski-Tinten, C. Uher, G. Wang, J.S. Wark, D. Zhu, D.A. Reis, Fourier-transform inelastic X-ray scattering from time- and momentum-dependent phonon-phonon correlations. Nat. Phys. **9**(12), 790–794 (2013)
21. S.L. Johnson, P. Beaud, E. Vorobeva, C.J. Milne, É.D. Murray, S. Fahy, G. Ingold, Directly observing squeezed phonon states with femtosecond X-ray diffraction. Phys. Rev. Lett. **102**(17), 175503 (2009)
22. M. Kozina, M. Trigo, M. Chollet, J.N. Clark, J.M. Glownia, A.C. Gossard, T. Henighan, M.P. Jiang, H. Lu, A. Majumdar, D. Zhu, D.A. Reis. Heterodyne X-ray diffuse scattering from coherent phonons. Struct. Dyn. **4**, 054305 (2017)
23. S.K. Sinha, Theory of inelastic X-ray scattering from condensed matter. J. Phys. Condensed Matt. **7511**, 7511–7523 (2001)

24. R. Santra, Concepts in X-ray physics. J. Phys. B Atomic, Molecular Opt. Phys. **42**(2), 023001 (2008)
25. G. Dixit, O. Vendrell, R. Santra, Imaging electronic quantum motion with light. Proc. Natl. Acad. Sci. **109**(29), 11636–11640 (2012)
26. G. Dixit, J.M. Slowik, R. Santra, Theory of time-resolved nonresonant X-ray scattering for imaging ultrafast coherent electron motion. Phys. Rev. A **89**(4), 043409 (2014)
27. B.W. Batterman, H. Cole, Dynamical diffraction of X-rays by perfect crystals. Rev. Modern Phys. **36**(3), 681 (1964)
28. B.E. Warren, *X-ray Diffraction* (Courier Corporation, North Chelmsford, 1990)

Chapter 3
Resonantly Bonded Semiconductors

Abstract Resonant bonded narrow-gap semiconductors have recently regained attention due to their unique structural properties and associated functional characteristics. In this chapter, I delve into the motivations behind investigating resonant bonded main group semiconductors, with a specific focus on SnSe, Bi_2Te_3, and Bi_2Se_3. These materials exhibit remarkable thermoelectric properties and also hold potential for electro-optic applications. Additionally, they serve as prototypical examples of topological semiconductors. By examining the fundamental physics underlying their demonstrated properties, I aim to provide some insights into the intriguing features of these materials.

The main group semiconductors I studied during my Ph.D. have been studied since 70 years ago. However, they keep giving us surprises. Some of them have, in the past decade, been found to be extremely good thermoelectric [1–8] and phase change materials [9, 10], and a lot of them appear in the search for prototypical topological materials. I introduce the two orthogonal aspects of these resonantly bonded narrow-gap semiconductors, both closely tied to functional properties. First, what is resonant bonding, and why does it make materials interesting? Second, what are the topological properties of these narrow-gap main group semiconductors?

3.1 Resonant Bonding

Two recent papers, namely, Lee et al. [11] and Lencer et al. [10], focus on resonantly bonded materials and their associated functionalities. These materials, including group IV–VI, V_2–VI_3, and V compounds, exhibit low thermal conductivity, which is a desirable characteristic for thermoelectric applications. Thermoelectrics feature poor thermal transport (phonon glass) but efficient electron conduction (electron crystal), enabling the effective conversion of heat into electricity. Additionally, IV–VI compound alloys serve as excellent phase change materials, displaying significant changes in optical and electrical properties upon crystallization. This

Y. Huang, *Towards the Optical Control of Resonantly Bonded Materials*, Springer Theses, https://doi.org/10.1007/978-3-031-42826-5_3

property allows for applications such as rewritable optical storage. In this section, we will delve into the structures, chemical bonding, interatomic interactions, and functional properties of these materials.

Orbital Hybridization, Bonding Coordination, and Structural Tunability

The low thermal conductivities of IV–VI materials compared with the III–V materials have been attributed to the fact that III–Vs have tetrahedral bonding, while IV–VIs have octahedral bonding. This difference in local coordination has the root in the s and p orbital hybridization. For example, in a lot of IV–VIs, the sp hybridization is small, and the s band is often lower in energy than the p band by multiple eV [12–14]. Therefore, one can consider mainly p electrons for valence states in IV–VIs. The average number of s and p electrons per atom in IV–VI and V semiconductors is above four, while III–V semiconductors possess on average four valence electrons per atom. A strong sp^3 hybridization is favorable for the latter and leads to tetrahedral bonding. On the contrary, a relatively pure p orbital leads to octahedral bonding, which is compatible with the natural wave function appearances of p_x, p_y, p_z. See Fig. 3.1(a).

Essentially, the term "resonant bond" means unsaturated covalent bonds (less than two electrons per bond). In IV–VI and V semiconductors with local octahedral coordination, there are, on average, three valence p electrons and six nearest neighbor bonds for each atom. The electrons are not enough to saturate the bonds. The ground state of a resonantly bonded system can be pictured as a superposition of symmetry equivalent saturated bond configurations [15]. See Fig. 3.1(b). The electrons are not localized in any of the saturated bond configurations.

For binary compounds composed of atoms A and B, Littlewood [16] defined two parameters based on which the phase diagram of main group resonantly bonded materials can be sorted out.

$$r_\sigma = r_p^A - r_p^B \tag{3.1}$$

$$r_\pi^{-1} = [(r_p^A - r_s^A) + (r_p^B - r_s^B)]^{-1} \tag{3.2}$$

r_s and r_p denote the valence radii of the s and p orbitals of the atom. The first parameter r_σ is a measure for the bond ionicity or the electronegativity difference. The second parameter r_π^{-1} describes the degree of covalency and measures the energetic splitting of s and p orbitals, which scales with the difference between s and p orbital radii. Larger r_π^{-1} leads to more saturated covalent bonds and a departure from pure p bonding. See Fig. 3.1 bottom and left schematic drawings where the bonding mechanism varies with the r_σ and r_π^{-1} coordinates. The starting point is a structure with predominant resonance bonding. The resonance character

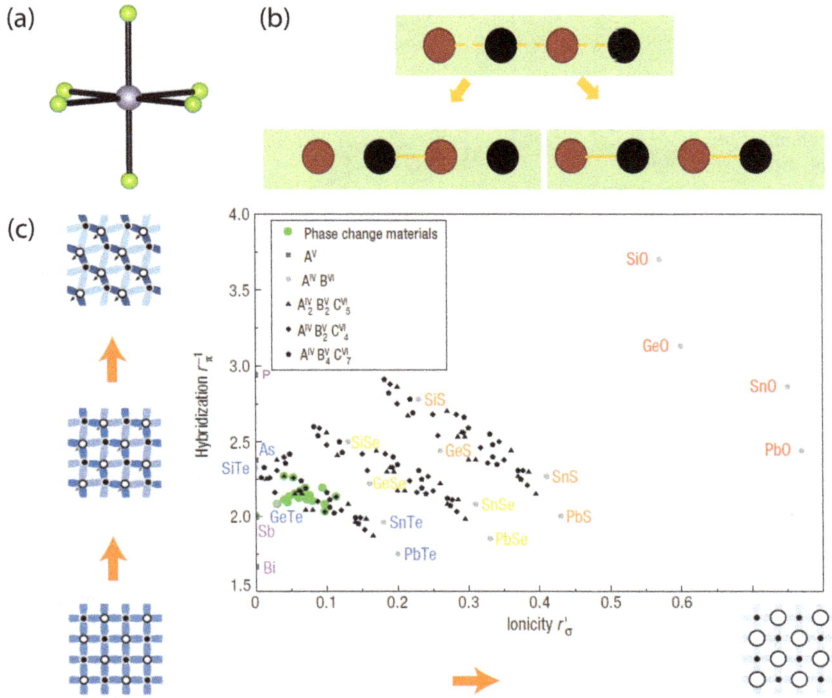

Fig. 3.1 (a) Gray and green atoms each represent an atomic species in a resonantly bonded material. The gray atom has six nearest neighbor bonds connecting to green atoms. (b) Unsaturated bonds of an undistorted 1D atom chain has a ground-state wave function that is a superposition of two saturated bond configurations. (c) A wide variety of materials is shown which differ in their tendency toward hybridization (covalency) r_π^{-1} and ionicity r_σ. Three types of system are shown: group V elements (squares), binary IV–VI compounds (gray circles), and ternary alloys with different compositions (triangles, diamonds, pentagons). Bands of oxides, sulfides, selenides, and tellurides are clearly discernible and are denoted by red, orange, yellow, and blue letters to label the corresponding binaries. The schematic drawing on the left shows that the structures deviate from the cubic as r_π^{-1} increases. The schematic drawing on the bottom shows that the electrons get more localized as r_σ increases. (Figure (c) is reproduced from Ref. [10] with permission from Springer Nature)

is weakened by increasing *sp* hybridization. See the left side inset. A high *sp* hybridization leads to distortions that favor more saturated covalent bonds and fewer of them (also reference Fig. 3.1(b)). On the other hand, larger r_σ means increased ionicity and leads to more localized electrons near ion cores. See the bottom side of the inset. The increase of both parameters r_σ and r_π^{-1} corresponds to increased charge localization and diminished resonant bonding and hence smaller electronic polarizability. Thus, strongly resonant bonding with highly delocalized valence electrons can only be found in the phase space where both r_σ and r_π^{-1} are small, as shown by the green area in Fig. 3.1(c), which is exactly the region of materials

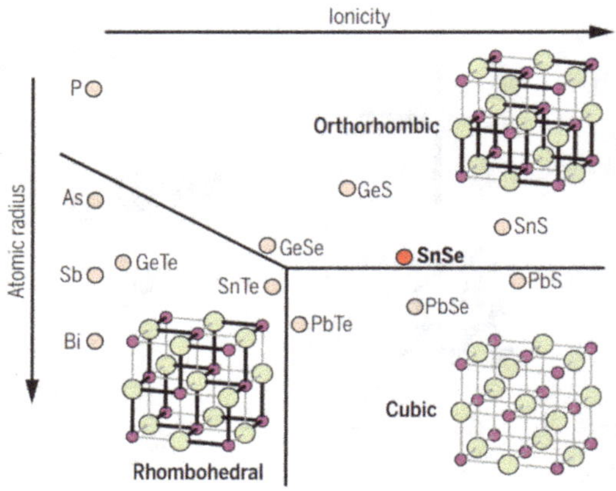

Fig. 3.2 For V and IV–VI semiconductors, depending on the ionicity and $s-p$ hybridization level (atomic radius), you can get cubic, orthorhombic, or rhombohedral structures. (The figure is reproduced from Ref. [17] with permission from the American Association for the Advancement of Science)

I study. The schematic drawing on the left shows that the structures deviate from the cubic as $s-p$ hybridization or r_π^{-1} increases. The schematic drawing on the bottom shows that the electrons get more localized and ionic bonds are formed as r_σ increases.

One of the interesting properties of resonant bonded materials is their structure tunability [16–18]. As shown in Fig. 3.2 for resonantly bonded IV–VI or V semiconductors and semimetals, there are several types of structural phases they can take. Note that the atomic radius and ionicity are another way to parameterize r_σ and r_π^{-1} above. The stability of these phases is sensitive to external parameters, including temperature, pressure [19], as well as stoichiometry [10].

The notion of resonant bonding is not limited to the IV–VI semiconductors. It expands to V_2–VI_3 and V materials where *sp* hybridization is weak and *s* band is located well below the *p* bands [18, 20]. In fact, their crystal structures are all derived from the rock salt lattice (NaCl-like). In Fig. 3.3(a) and (b), I pick an example from IV–VI (PbTe) as well as an example from V_2–VI_3 (Bi_2Te_3). In both materials, the parent structure is identifiable.

Long-Range Interatomic Interactions, Soft Phonons, and Low Thermoconductivity

In Ref. [11], it is pointed out that resonant bonding leads to long-range forces, which is a general statement. In Fig. 3.3(a) and (b), I show that the 1st, 4th, 8th, and 14th nearest neighbor atoms are aligned along $\langle 100 \rangle$. Figure 3.3(c–d) shows the interatomic force constants (IFC) versus atomic distance.[1] In PbTe (see Fig. 3.3(c)),

[1] To compare different materials regardless of crystal structures and bonding stiffness, traces of IFC tensors are normalized by the trace values of the self-interacting IFC tensor.

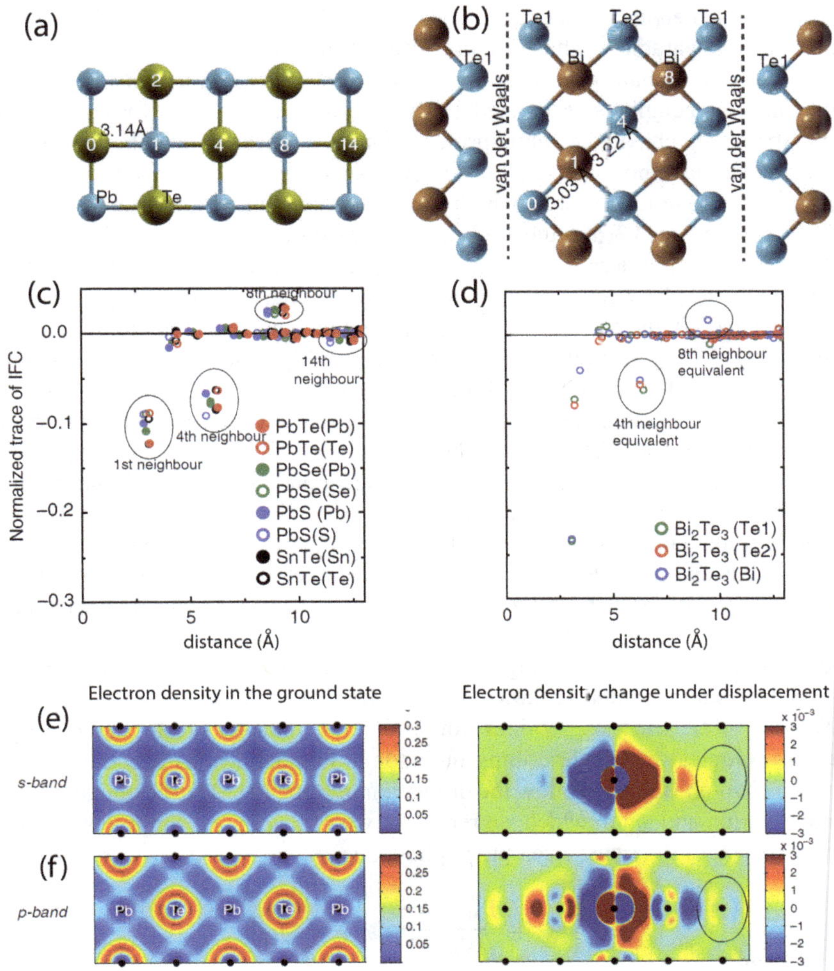

Fig. 3.3 (a) Cubic resonant bonding network of a IV–VI semiconductor representative PbTe. The atoms along ⟨100⟩ are marked with nearest neighbor indices 1st, 4th, 8th, and 14th. (b) Rhombohedral V_2–VI_3 representative Bi_2Te_3, which is just a slight distortion from a rock salt within the quintuple layers. (c–d) Normalized trace of interatomic force constant tensors versus atomic distances. (c) Lead chalcogenides and SnTe (group IV–VI). (d) Bi_2Te_3 (group V_2–VI_3). (e–f) Electron density distribution change by the displacement of the center atom in PbTe. The plot is on (100) plane, and each black dot represents an atom. The unit is $Å^{-3}$. (e) is a charge projection onto the s orbitals, and (f) is a projection on p orbitals. The long-range interactions are mainly mediated by the p orbitals. (Figures are reproduced from Ref. [11] with permission from Springer Nature)

fourth nearest neighbors, separated by 6 Å(Pb-Pb and Te-Te), have interactions which are comparable to those of first nearest neighbors, spaced 3 Å apart, and are much stronger than second and third nearest neighbor interactions. In addition, eighth nearest neighbors, separated by 9 Å, have even positive force constants, giving them the behavior of "anti-springs." Fourteenth nearest neighbors, separated by 12 Å, also have non-negligible force constants. This is certainly not attributed to a Coulomb long-range force, which would be monotonic in the atomic distance. In Bi_2Te_3 (see Fig. 3.3(d)), the IFC dependence on atomic distance is similar to PbTe, though slightly less significant. The major difference is that there is no long-range interaction between atoms connected by the van der Waals-type bonds. In Fig. 3.3(e) and (f), the electron density perturbed by central atomic motion in PbTe is long-ranged for p orbital projected density (see Fig. 3.3(e)) but much less for s orbital projected density (see Fig. 3.3(f)). The p orbital resonant bonding is clearly associated with the long-range force, which by the way cannot be captured by simplistic force models (shell model, bond charge model, etc.).

One of the interesting properties of resonant bonded materials is their lattice instability. As shown in Fig. 3.4(a) for resonantly bonded IV–VI or V semiconductors and semimetals, there are several types of structural phases they can take. The atomic radius and ionicity are another way to parameterize r_σ and r_π^{-1}. The lattice instabilities are often characterized by the zone center TO mode frequency ω_{TO}. In Fig. 3.4(b), it is shown that for the 1D chain, the softening of zone center TO mode increases as the force along the chain becomes longer-ranged. This is indeed observed in real materials, as shown in Fig. 3.4(c) for IV–VI and d for V_2–VI_3.

We can obtain the expressions for ω_{TO} in a binary crystal, based on the phenomenological model describing interatomic interactions, as introduced in the previous chapter. Below, I follow the derivation in Refs. [21, 22]. The binary crystal contains cation and anion which we represent with subscripts c and a, respectively. The frequency of the transverse optical phonon is

$$\omega_{\mathrm{TO}}^2 = \frac{e}{\mu \mathrm{V}_0}(\frac{4\pi}{3} Z_a Z_c - \kappa), \tag{3.3}$$

$$\kappa = \frac{\Omega_0^2}{4e^2} \sum_{\mathbf{G},\mathbf{G}',\mathbf{G}''\neq 0} \epsilon^*_{\mathbf{G},\mathbf{G}'} \chi_{\mathbf{G},\mathbf{G}''} \mathrm{V}_c^*(\mathbf{G}') \cdot \mathrm{V}_a(\mathbf{G}''), \tag{3.4}$$

and

$$\mathrm{V}_\alpha(\mathbf{G}) = \sum_{\mathbf{G}'\neq 0} \epsilon^{-1}_{\mathbf{G},\mathbf{G}'} v_\alpha(\mathbf{G}')\mathbf{G}' e^{-i\mathbf{G}'\cdot \mathbf{r}_\alpha}, \tag{3.5}$$

where μ is the reduced mass, Ω_0 is the unit cell volume, and α represents either cation or anion. The term κ corresponds to forces between ions mediated by electron redistribution. $v_\alpha(\mathbf{G})$ is the Fourier transform of the bare electron-ion pseudopotential. In the last chapter, I did not talk about the form of $\chi(\mathbf{r}, \mathbf{r}')$. In

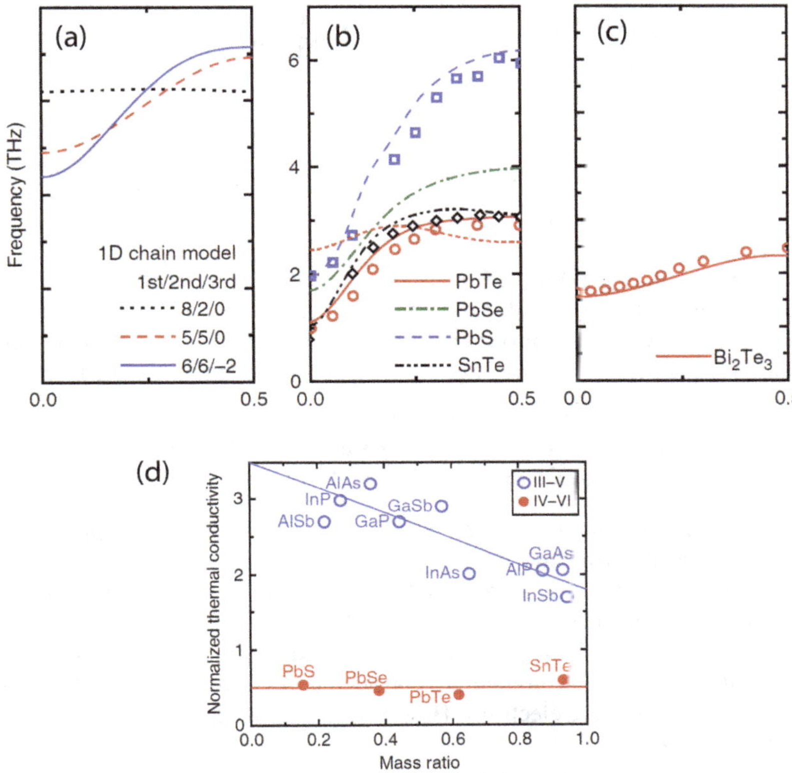

Fig. 3.4 Near-ferroelectric behavior due to resonant bonding. (a) Optical phonon dispersion in a model 1D atomic chain, showing the softening of the optical phonons due to the long-ranged interactions. Three numbers in the legend represent the relative interaction strengths of first, second, and third nearest neighbors in the 1D chain. (b) Soft TO phonon modes along the trigonal direction for lead chalcogenides and SnTe. (c) Soft TO phonon modes along the trigonal direction for Bi_2Te_3. Mode frequencies are calculated based on first principles. Lines and circles are calculation and experimental data, respectively. (d) Normalized thermal conductivities of binary III–V and IV–VI compounds at 300 K. The solid lines are for a guide to the eye. (Figures are reproduced from Ref. [11] with permission from Springer Nature)

a periodic crystal, the perturbations and induced charge density both occur with the same periodicity as the lattice. $\chi(\mathbf{q}, \mathbf{q}')$ on its Fourier domain can be assumed to be nonzero only when $\mathbf{q} = \mathbf{G}$ ($\mathbf{q}' = \mathbf{G}'$). We have dielectric constant $\epsilon_{\mathbf{G},\mathbf{G}'} = \delta_{\mathbf{G},\mathbf{G}'} - \frac{4e^2}{G^2}\chi_{\mathbf{G},\mathbf{G}'}$, and V is the gradient of screened electron-ion potential, expressed in the Fourier space. The authors went on to obtain an analytical expression for κ, based on the assumption that off-diagonal elements and $\mathbf{G} = 0$ term in $\epsilon(\mathbf{G}, \mathbf{G}')$ can be ignored. The off-diagonal elements of linear response χ or the dielectric function ϵ reflects the nonuniformity of the charge density within one unit cell and provides Umklapp scattering. In this case,

$$\kappa = \frac{\Omega_0^2}{12e^2} \sum_{\mathbf{G}\neq 0} \tilde{S}(\mathbf{G}) G^2 [v_S^2(\mathbf{G}) - v_A^2(\mathbf{G})] \frac{G^2}{4e^2} \epsilon_{\mathbf{G}} (1 - \epsilon_{\mathbf{G}}), \tag{3.6}$$

and

$$\tilde{S}(\mathbf{G}) = \sum_{\mathbf{G}\text{ and symm. equiv.}} \exp\left[i\mathbf{G} \cdot (\boldsymbol{r}_{\text{cation}} - \boldsymbol{r}_{\text{anion}})\right], \tag{3.7}$$

where $v_S(\mathbf{G}) = 0.5[v_a(\mathbf{G}) + v_c(\mathbf{G})]$, $v_A(\mathbf{G}) = 0.5[v_a(\mathbf{G}) - v_c(\mathbf{G})]$, and they are, respectively, the symmetric and antisymmetric form factors. As remarked in [23], the main advantage of this approach is to demonstrate the physical origin of soft TO phonons. We see that for κ to be large, the two properties below can be helpful:

(1) Rock salt structure, which is considered a typical ionic structure
(2) Covalency, which means $v_S(\mathbf{G})$ is large compared to $v_A(\mathbf{G})$ for the leading order of $\mathbf{G}$, leading to a large $[v_S^2(\mathbf{G}) - v_A^2(\mathbf{G})]$ factor

In zincblende or diamond lattices, $\tilde{S}(\mathbf{G}) = 0$ for $\mathbf{G} \in \{111\}$ [reciprocal lattice vector (111) and its symmetry equivalents], for which $v_S(\mathbf{G})$ is the largest. Since $v_S(\mathbf{G})$ decreases steeply with increasing $\mathbf{G}$, κ is small in this case. This implies that the covalent semiconductors such as Si or GaAs, which do obey condition (2) but not (1), are also stable to lattice distortions. The condensation of the TO mode under electron-lattice interaction is sometimes considered a Peierls distortion, of which the instability comes from the electrons. However, the above derivation did not invoke electronic gap opening.

The same pseudopotential approach as presented above can also be used to calculate the Born effective charges and dielectric constants, which I touched on in the previous chapter. In Ref. [24], Littlewood built a one-dimensional diatomic model that describes the ground-state charge density under given lattice distortion and ignoring screening. The lattice distortion gives rise to a change in bonding charge proportional to v_S/v_A, the ratio between the symmetric part of the pseudopotential and the antisymmetric part. A model generalized to a three-dimensional binary crystal [25, 26] shows that the Born effective charge is proportional to $v_S(111)/v_A(111)$ to the low order. Hence, the low TO frequency, large Born effective charges, as well as high dielectric constants are associated with both the ionic structure and high covalency of the bonding. This phenomenological view on the nature of resonantly bonded rock salt structure materials fits nicely with the ab initio DFT results in Ref. [11].

Long-range force leads to large anharmonicity [14, 27] that enhances the multi-phonon scattering process, and low-lying optical modes increase the phase space for phonon-phonon scattering. Both effects lead to low thermoconductivity and thus a high figure of merit of thermoelectricity [11]. As shown in Fig. 3.4(d), the normalized thermal conductivity of resonantly bonded IV–VI semiconductors is much lower than III–V semiconductors.

Phase Change Materials

To get good phase change materials that change quickly between crystalline and non-crystalline phases of matter, one needs large atomic fluctuations or large structural entropy that facilitates phase transitions. Resonantly bonded materials, in general, have flat anharmonic atomic potentials due to long-range forces. Thus, many resonant bonded materials and their alloys are good phase change materials. The authors of Ref. [10] point out that to maximize atomic fluctuations, a slightly distorted resonant bond is desired. See Fig. 3.5. If the lattice is of high symmetry, but there is not much electron-lattice interaction, or the crystal is highly ionic and charges just localize around atoms, we do not have a flat potential (left). When the structure is strongly distorted, we also do not have a flat potential (right). Resonantly bonded materials, with a high-symmetry parent structure, covalent bonding, and a large Born effective charge due to long-range interactions (or a large renormalization of interatomic interactions due to a large linear response of electrons), have flat potentials (middle) and hence are good phase change materials.

SnSe as a Representative of IV–VI

SnSe represents rock salt-like IV–VI compounds that host several lattice instabilities associated with their nearly cubic resonant bonding network and possess properties I mention for general resonantly bonded materials. I will provide a detailed discussion on the different phases of SnSe in Chap. 5.

Below, I will elaborate on lone pairs, whose effect on lattice instability can be largely understood from a local Jahn-Teller-like interaction (see the first section of Chap. 2). Specifically, we look at lone pairs in IV–VI semiconductors, taking SnSe as an example. In Fig. 3.6(a), the calculated orbital projected electronic density of states (DOS) of the low-T structure shows that bonding states near the Fermi level

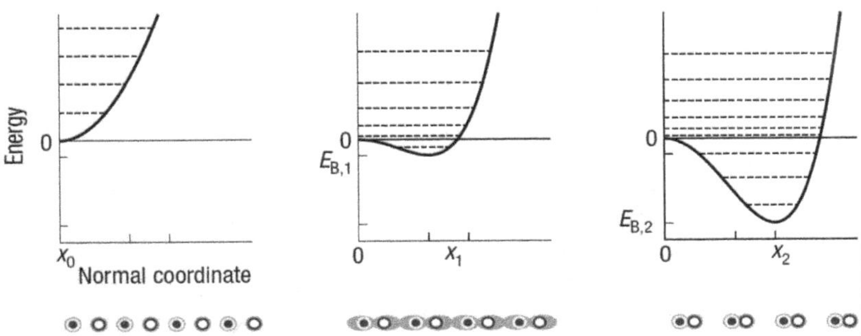

Fig. 3.5 Schematic model potentials in the case of local distortions. (The figure is reproduced from Ref. [10] with permission from Springer Nature)

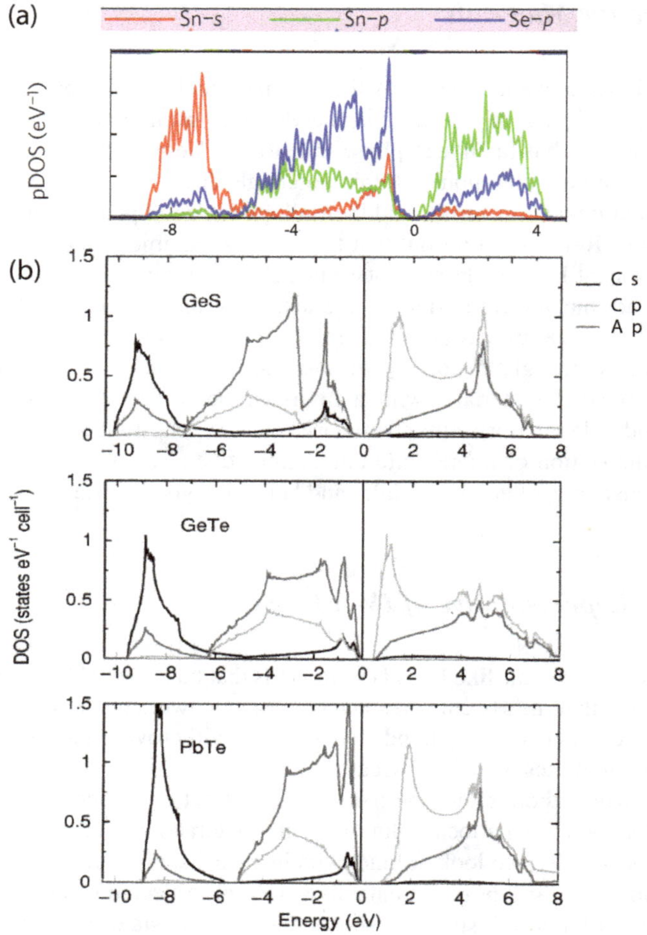

Fig. 3.6 (a) Orbital projected electronic density of states in the Pnma phase SnSe. (b) Orbital resolved densities of states for GeS, GeTe, and PbTe in the rock salt structure. The origin on the energy axis is the top of the valence band. A represents anion; C represents cation. (Figure (a) is reproduced from Ref. [14] with permission from Springer Nature. Figure (b) is reproduced from Ref. [13] with permission from the American Physical Society)

are dominated by Se p states, which form broadband (6 eV wide) hybridized with the Sn s band and mediated by the Sn p states. The Sn s states are mainly confined 8 eV below the Fermi level, but also show a peak just below the top of the valence band, from hybridization with p states. This DOS peak is Sn 5s lone pair formation from hybridization with Se p. In Fig. 3.6(b) GeS, GeTe, and PbTe, we also see the formation of lone pair (mixed s and p orbitals), which corresponds to a sharp DOS right below the Fermi level. The stereo-chemical activity of lone pairs depends on the bandwidths of and the separation between the cation p, anion p, and cation s

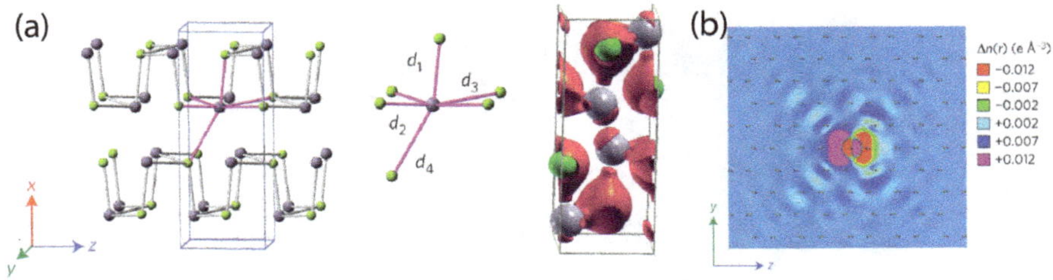

Fig. 3.7 (a) Crystal structure of room temperature SnSe phase. The structure is broken into bilayers inside each unit cell. Sn atoms are in gray and Se in green. The valence electron density (isosurface at 0.26 Å^{-3}) is computed from first principles. The lone pairs are "caps" around the gray Sn atoms, which are located between bilayers. (b) The map of electron density change induced by TO phonon displacement (Sn atom 0.02 Å along **c**). Oscillations along diagonals show long-range interactions in the **b**–**c** plane along the resonant bonding direction. (Figures are reproduced from Ref. [14] with permission from Springer Nature)

bands. In principle, the stronger the mixing of cation s state as well as cation p state with the anion p state the stronger the lattice instability can be induced by lone pairs—one can refer to Eq. 2.1. If the mixing is over a large bandwidth, the lone pair will be diffused in energy and space. Otherwise, it can be pretty localized.

The local (distorted) octahedral coordination of SnSe can be seen from Fig. 3.7(a). This octahedron containing $d_1 - d_4$ can be thought of as a building block of its parent rock salt under distortion. Figure 3.7(a) also provides a visualization of the lone pair, the little red "caps" at the corner of the Sn (gray) atoms. The simulations show that upon TO displacement, there is a long-range charge perturbation in SnSe along the resonant bonding direction $\langle 011 \rangle$, consistent with our discussion in the previous section.

Bi_2Te_3 *and* Bi_2Se_3 *as Representatives of* V_2VI_3

Bi_2Te_3 has a rhombohedral structure that can be understood as a deformed rock salt structure. The structure is a stacking of quintuple layers, each containing five resonantly bonded layers of atoms ($Te^{(1)}$–Bi–$Te^{(2)}$–Bi–$Te^{(1)}$). Each quintuple layer is separated from the next by weak Te1–Te1 van der Waals interactions. The structural deviation from the exact rock salt structure within the quintuple layer is small. The lengths of the strongest bond ($Te^{(1)}$–Bi), 3.03 Å, and that of the second strongest bond (Bi–$Te^{(2)}$), 3.22 Å, are similar. In addition, the angles of Te1–Bi–Te2 and Bi–Te2–Bi are 174.6° and 180°, respectively, close to exactly 180° bond angles in the rock salt structure; see Fig. 3.8. Owing to this small structural distortion, the resonant bonding exists in a weakened form in Bi_2Te_3.

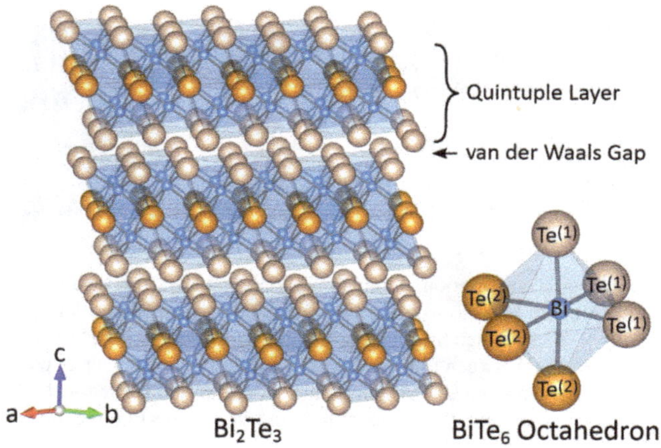

Fig. 3.8 Layered structure of Bi_2Te_3. Bismuth atoms are octahedrally coordinated by tellurium atoms on two inequivalent sites. Each quintuple layer is separated by weak van der Waals bonding between adjacent Te(1) layers. (The figure is reproduced from Ref. [28] with permission from John Wiley and Sons)

3.2 Topological Phases

Topological materials represent a newly discovered quantum state of matter which is characterized by peculiar edge or surface states that show up due to a topological character of the bulk wave functions. When distinguishing the phases of materials, apart from the perspective of free energy and symmetry breaking, there is another important perspective to look at phase transitions, the topology of the Hilbert space.

It turns out that some of the resonantly bonded semiconductors are in symmetry-protected topological phases [29–31], including the three materials I study in this thesis, which are SnSe (in its cubic phase) [32], Bi_2Se_3, and Bi_2Te_3 [33]. Among them, Bi_2Se_3 and Bi_2Te_3 are topological insulators (TI) protected by time-reversal symmetry, while SnSe (cubic) are topological crystalline insulators (TCI) protected by crystalline symmetry. Both TIs and TCIs are defined with topological invariants encoded by the wave functions of Bloch electrons. The symmetry protection is crucial for these topological states, as the removal of symmetry will allow the electronic structures, whose topology is characterized by topological invariants, to deform adiabatically into a trivial atomic limit.

Since time-reversal symmetry is a local symmetry and thus less complicated than a crystalline symmetry, we start by talking about topological insulators (TI), which are states protected by time-reversal symmetry, before we talk about topological crystalline insulators (TCI), which are states protected by non-local crystalline symmetry. I first outline how to quantify the topology of a TI using a topological invariant from the perspective of fermion parity change.

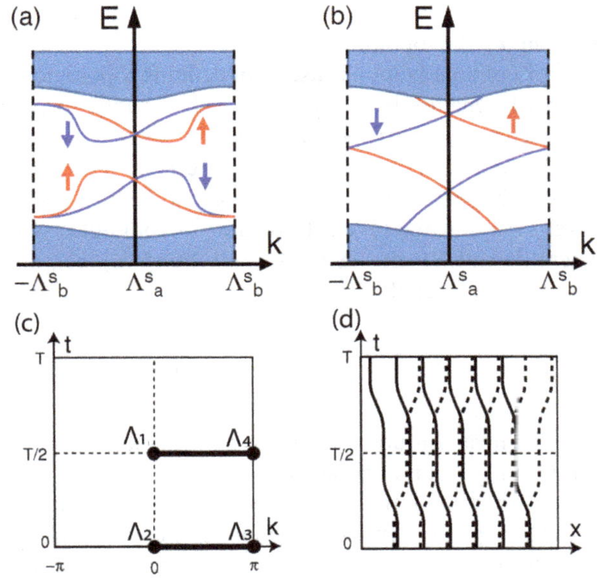

Fig. 3.9 (a–b) Schematic pictures of the surface states between two surface TRIMs, Λ^s_a and Λ^s_b, for (a) topologically trivial and (b) topologically nontrivial cases. In the latter, the Kramers doublet switch partners between different TRIMs. The shaded regions represent the bulk continuum states. (c) A system with both k and t obeying periodic boundary conditions has its topological variant depending on four Λ taking (k, t) values $\Lambda_1 = (0, 0)$, $\Lambda_2 = (0, T/2)$, $\Lambda_3 = (\pi, 0)$, and $\Lambda_4 = (\pi, T/2)$. (d) A schematic showing the evolution of the Wannier centers of a Kramers pair (the solid line and the broken line). (Figures (a) and (b) are reproduced from Ref. [34]. ©(2013) The Physical Society of Japan. Figures (c) and (d) are reproduced from Ref. [35] with permission from the American Physical Society)

In quantum mechanics, the Kramers degeneracy theorem states that for every energy eigenstate of a time-reversal symmetric system with half-integer total spin, there is at least one more orthogonal eigenstate with the same energy. In Fig. 3.9(a)–(b), I show two Kramers pairs, each containing states of opposite spins. The Kramers pairs are degenerate at the high time-reversal invariant momenta (TRIM, $\mathbf{k} = -\mathbf{k}$ mod ($\mathbf{G}$), e.g., 0, π or Λ^s_a, Λ^s_b in Fig. 3.9(a)–(b)). In the topologically nontrivial phase, the Kramers pairs at $k = \pi$ are not the same as those at $k = 0$. In the topologically trivial case, however, the pairs are the same. As a consequence, in the topological case, there is an odd number of levels crossing zero energy, while in the trivial cases, there is an even number of them. A change from a topologically trivial phase to a nontrivial phase needs to change the number of bands crossing the Fermi level between TRIMs $k = 0$ and $k = \pi$ (Fermi level can be anywhere in Fig. 3.9(a)–(b)), by an odd number. This will inevitably involve closing the gap. To define a topological invariant, we would want to define a quantity that considers the whole k space, and it should not depend on where the bands cross the Fermi level, because we can always deform the band adiabatically and move the crossing point around without changing the topology.

The modern theory of polarization is useful in defining topological invariant [36], and the fermion parity change is a reflection of the change in polarization. The polarization **P** of a certain band n is defined as the center of charge of Wannier states [37]. It can be calculated by integrating the Berry connection of the occupied states over the Brillouin zone. Following the pedagogical Ref. [34], for a 1D two-band time-reversal system (time-reversal operator Θ commutes with Hamiltonian H) with a Kramers pair of Bloch functions $|u_1(k)\rangle$ and $|u_2(k)\rangle$, one can define a difference polarization between the Kramers pairs,

$$P_\theta = \int_{-\pi}^{\pi} dk \frac{-i}{2\pi}[\langle u_1(k)|\nabla_k|u_1(k)\rangle - \langle u_2(k)|\nabla_k|u_2(k)\rangle]. \tag{3.8}$$

Since the topology is invariant upon adiabatic deformation of the electron bands, we can generically think about an adiabatic evolution of the electron energy level that is periodic (returns to the initial state after a complete cycle of evolution) and later assign the adiabatic parameter with the momentum. In crystals, k indeed satisfy the periodic boundary condition. In Ref. [34], it is assumed that the 1D two-band model is periodic in time with periodicity $T = 2\pi$ and is time-reversal symmetric at $t = 0$ and $t = \pi$. Thus, at these time points, according to the Kramers theorem, $\Theta|u_2(k)\rangle = e^{-i\chi(k)}|u_1(-k)\rangle$, and $\Theta|u_1(k)\rangle = -e^{-i\chi(k)}|u_2(-k)\rangle$ (the minus sign of $-e^{-i\chi(k)}$ before the exponent is because $\Theta^2 = -1$ for particles of half-integer spin). A matrix w can be defined to describe the effect of time-reversal operator Θ on states:

$$w_{\alpha\beta}(k) = \langle u_{\alpha,-k}|\Theta|u_{\beta,k}\rangle, \tag{3.9}$$

and therefore,

$$w(k) = \begin{pmatrix} 0 & e^{-i\chi(k)} \\ -e^{-i\chi(-k)} & 0 \end{pmatrix}. \tag{3.10}$$

After some algebra, it can be shown that

$$P_\theta = \frac{1}{i\pi} \log\left(\frac{\sqrt{w_{12}(k=0)^2}}{w_{12}(k=0)} \frac{w_{12}(k=\pi)}{\sqrt{w_{12}(k=\pi)^2}}\right) \tag{3.11}$$

and P_θ can only take values of ± 1. The value of P_θ is not meaningful by itself since it can be changed via a gauge transformation. Its physical meaning is the difference between the Kramers pairs' Wannier center shifts over the course of a half period ($t = 0$ to $t = \pi$). In general, the Hilbert space ($|u_n(k, t)\rangle$) can be classified into two groups depending on the difference in P_θ at $t = 0$ and $t = \pi$. A quantity Δ defined as

$$\Delta = P_\theta(T/2) - P_\theta(0) \tag{3.12}$$

and $\Delta = 0,1 \pmod 2$ is a Z_2 topological invariant that characterizes the Hilbert space. There is also the relation

$$(-1)^{\Delta} = \prod_{i=1}^{4} \frac{w_{12}(\Lambda_i)}{\sqrt{w_{12}(\Lambda_i)^2}}. \tag{3.13}$$

The Λ can take the pairs of (k, t) values $\Lambda_1 = (0, 0)$, $\Lambda_2 = (0, T/2)$, $\Lambda_3 = (\pi, 0)$, and $\Lambda_4 = (\pi, T/2)$ (see Fig. 3.9(c)). If $\Delta = 1 \pmod 2$, it physically means that in the adiabatic evolution from $t = 0$ to π, the fermion parity changes. One can think of this as the line Λ_1–Λ_4 resembling Fig. 3.9(b) and the line Λ_2–Λ_3 resembling Fig. 3.9(a). To see this adiabatic evolution in position space, one can plot the Wannier center motion as a function of time, in Fig. 3.9(d). Between $t = 0$ and $t = \frac{T}{2}$, the Wannier states switch their partners.

This quantity Δ is defined for the bulk ground state and, due to the bulk boundary correspondence, leads to edge modes that are gapless. Usually, symmetry-protected topology is defined with two notions: (1) the bulk ground state cannot be adiabatically connected to the atomic limit without breaking the relevant symmetry, and (2) the gapless boundary modes can *only* be gapped by breaking the relevant symmetry. I have outlined how to find a topological invariant that can serve as a criterion for (1), and I do not dive into (2), which, depending on the respective symmetry, can be complicated in interacting systems.

The definition of topological invariant Δ requires only the system to be periodic. A crystal has periodic boundary condition in k, and with time-reversal symmetry, a similar definition for Δ will apply to multi-dimensional systems and multiple bands (Δ needs to be the sum over *all* occupied bands in a bulk insulator and *all* TRIMs) [34, 35]. For a 2D system, Δ can be rewritten as

$$C = \sum_n \frac{-i}{2\pi} \int_{BZ} d^2\boldsymbol{k} \cdot \nabla_k \times \mathrm{Tr}(\boldsymbol{a}), \tag{3.14}$$

and $\Delta = C \pmod 2$. $\mathrm{Tr}(\cdot)$ means taking the trace of the matrix. $\boldsymbol{a}_{\alpha\beta}(k) = -i\langle u_{\alpha,k}|\nabla_k|u_{\beta,k}\rangle$, and n denotes all occupied bands. The physical meaning of C written this way is the net number of charge centers crossing a given position x as one wraps around a Brillouin zone along k_y [38]. C is often called a Chern number, which must take integer values (at least in non-interacting systems). In the special case where the system happens to have both time-reversal and spatial inversion symmetry, the Z_2 topological invariant Δ is only related to the product of (inversion) parity eigenvalues (of occupied bands) at TRIMs [39].

The definition of topological invariant can be generalized to the TCI as well. For TCI with mirror symmetry M ($[M, H]$=0), for example, the Hilbert space is partitioned into two groups. The first group is the M eigenstates with eigenvalue i; the other is the M eigenstates with eigenvalue $-i$. Note that $M^2 = -1$ because $M = PC_2$ is a multiplication of inversion and twofold rotation and spinors produce a phase $e^{i\pi}$ upon a 2π rotation in M^2. The difference of the Chern numbers of these two subspaces divided by 2, $C_M = (C_{+i} - C_{-i})/2$, can denote a "mirror polarization" that characterizes the topology. Note that gapless surface/edge modes in a TCI require a certain termination of the material due to the non-local nature of crystalline symmetry.

Below, we look at the specific materials that will be studied in this thesis.

In the spirit of modern condensed matter physics, Hamiltonians describing the lower-energy physics can be summarized based on its symmetry. One does not need to know the details of the Hamiltonian (e.g., whether it is nearest neighbor coupling, next nearest neighbor coupling, or further neighbors) to predict important structures of electron bands. A good text on this is Ref. [38] Chapter 1.

The first step is to identify the low-energy electron bands, starting with a free electron or tight-binding approximation, and then add in the crystal field effects and spin-orbit couplings. A good reference for this is Ref. [40] and, specifically for the materials studied in the thesis, Refs. [33, 41].

For IV–VI crystals in their cubic phase, the direct bandgap occurs at the L points instead of zone center Γ. At L point, the p bands from cation and anion have different parities, thus the p orbitals from the cation site (conduction band) and anion site (valence band) are not mixed, while at Γ, both valence band and conduction band derived from p orbitals are odd, thus electron bands are not majorly from either the cation or the anion. The little group of L in a cubic rock salt structure of IV–VI semiconductor is D_{3d}. One can look up its full table of characters; however, I only summarize relevant irreducible representations of lower energy bands in Table 3.1. The states Γ_4^+, Γ_5^+ and Γ_4^-, Γ_5^- will split off from the $\Gamma_6^+(L_3^+)$ and $\Gamma_6^-(L_3^-)$ states under spin-orbit coupling (SOC). And we expect them to lie above the Γ_6^+, Γ_6^- states just as in atomic physics $p_{\frac{3}{2}}$ is higher than $p_{\frac{1}{2}}$. Note that $\Gamma_6^{+/-}$ states are like spinors with two components. The low-energy physics really comes from $\Gamma_6^+(L_3^+)$ and $\Gamma_6^-(L_3^-)$. The presence of the s levels of IV cation below the p levels of VI anion lifts up the L valence bands as a result of the $s-p$ admixture, which is forbidden at Γ zone center, for the sake of parities. Depending on the $s-p$ mixture, in some IV–VI semiconductors like PbTe, the conduction band minimum is formed by the $\Gamma_6^-(L_3^-)$ state and the valence band edge has the $\Gamma_6^+(L_3^+)$, and in some others like SnTe, this ordering is reversed, which we will see is related to the topological surface states. The reversal happens at an intermediate alloying of PbTe and SnTe, which has been documented since 1966 [42].

Table 3.1 Irreducible representations at L point for rock salt IV–VI crystals

I.R. (w/o SOC)	L_1^+	L_3^+	L_2^-	L_3^-
I.R. (w/ SOC)	$\Gamma_6^+(L_1^+)$	$\Gamma_6^+(L_3^+) + \Gamma_4^+ + \Gamma_5^+$	$\Gamma_6^-(L_2^-)$	$\Gamma_6^-(L_3^-) + \Gamma_4^- + \Gamma_5^-$
Component	s	d-like near anion, p-like near cation	p	p
Wave function	Spherical	Pseudovector or e_g	p_x	p_y, p_z
Degeneracy of $\Gamma_6^{+/-}$	1	2	1	2
Parity	Even	Even	Odd	Odd

Next, we shall write down the effective Hamiltonian at low energies. The low energy states consist of two Kramers pairs of opposite parity. As explained, these opposite parity states originate from cation and anion, respectively, at L point. Each Kramers pair is denoted by $|\psi_\alpha^+\rangle$ and $|\psi_\alpha^-\rangle$. The two members of a Kramers doublet have opposite total angular momenta $J_z = \pm\frac{\hbar}{2}$.[2] There are, therefore, a total of four states to consider. One can take the route as in Ref. [43] to obtain the Hamiltonian, which is a standard $k \cdot p$ perturbation (including spin-orbit coupling). Alternatively, one can make sure that the Hamiltonian has the required symmetries and expands it to the low orders in k. The two routes will yield identical forms. The symmetries of the Hamiltonian include spatial inversion P, threefold rotation around (111) C_3 at L, and mirror symmetry for (1–10) plane M, as well as time-reversal Θ. If we define principal axes to be $\hat{k}_x = (1-10)$, $\hat{k}_y = (1-12)$, $\hat{k}_z = (111)$, $P = I \otimes \tau_z$ (spatial inversion), $C_3 = \exp(i\frac{\pi}{3}\sigma_z) \otimes I$ (threefold rotation), $M = -i\sigma_x \otimes I$ (mirror reflection), and $\Theta = iK\sigma_y \otimes I$ (time reversal), where K is the complex conjugation operator. Here, τ represents the orbital space, and σ is the spinor space. $M = PC_2$, the product of inversion and twofold rotation, and since the orbits are already defined as the parity eigenstates, M only has its effect on the spinor through the π rotation. Note that $M = -i\sigma_x \otimes I$ is only the case when the C_3 (111) is in the mirror plane (1–10); if it is perpendicular to the mirror plane, then $M = \sigma_z \otimes I$.

The Hamiltonian at L is then [30, 44]

$$H(k) = m\tau_z + v\tau_x(k_x\sigma_y - k_y\sigma_x) + v_z k_z \tau_y \tag{3.15}$$

It takes the form of a massive Dirac fermion in 3 + 1 dimension.

For Bi_2Se_3 at zone center Γ, its low-energy orbitals are $(|P_z^+,\uparrow\rangle, |P_z^-,\uparrow\rangle, |P_z^+,\downarrow\rangle, |P_z^-,\downarrow\rangle)$, according to Ref. [33]. The two positive parity states with a + superscript are mainly from Bi, and the two negative parity states from Se. The symmetry operators are $\Theta = iK\sigma_y \otimes I$ (time reversal), $C_3 = \exp(i\frac{\pi}{3}\sigma_z) \otimes I$ (threefold rotation), and $P = I \otimes \tau_z$ (spatial inversion). Here, τ represents the orbital space, and σ is the spin. The Hamiltonian at L is then

$$H(k) = \epsilon_0(k) + m(k)\tau_z + A_1 k_z \tau_x \sigma_z + A_2 k_x \tau_x \sigma_x + A_2 k_y \tau_x \sigma_y \tag{3.16}$$

It also takes the form of a massive Dirac fermion in 3+1 dimension [33]. Dirac Γ matrices are $\Gamma^1 = \tau_x \otimes \sigma_x$, $\Gamma^2 = \tau_x \otimes \sigma_y$, $\Gamma^3 = \tau_x \otimes \sigma_z$, $\Gamma^4 = \sigma_y \otimes I$, and $\Gamma^4 = \sigma_z \otimes I$, and the Hamiltonians for both IV–VI rock salt and Bi_2X_3 ($X \in \{Te, Se\}$) in the four-band model take the form

$$H(k) = d_0(k)I + \sum_{a=1}^{5} d_a(k)\Gamma^a \tag{3.17}$$

[2] Although it is a two-component spinor, it is not the "electron spin."

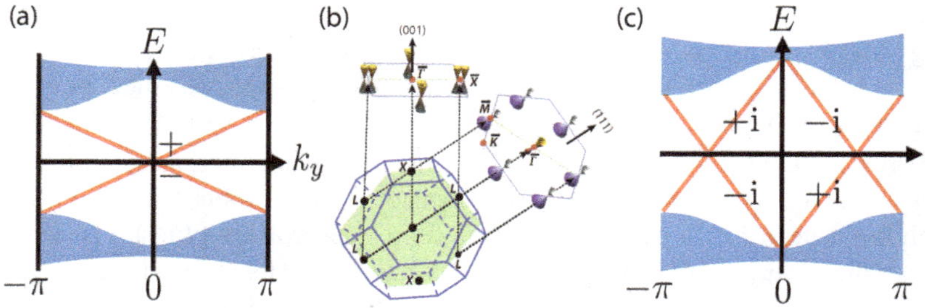

Fig. 3.10 Band structures with band inversion happening at an odd number of TRIM versus an even number of TRIM. (a) Edge states in TI where the Z_2 invariant is 1 and TCI with an odd number of TRIM band inversions (mirror Chern number C_M is odd). (b) IV-VI rocksalt 3D Brillouin zone (BZ), with (001) and (111) surface BZ, as well as illustrations of Dirac cones. (c) An even mirror Chern number C_M will enforce the presence of two chiral left edge states and two chiral right edge states along mirror symmetric lines on the M-invariant surface Brillouin zone. (Figures (a, c) are reproduced from Ref. [38] with permission from Springer Cham. Figure (b) is reproduced from [45] with permission from the American Physical Society)

For systems with both time-reversal and spatial inversion symmetry, the Z_2 invariant is only related to the $\text{sgn}[d_5(\Lambda_i)]$ (or "band inversion") at TRIMs.

When the system becomes a TI, the band order flips only at the Γ point in the 2D Brillouin zone (BZ) and realizes the nontrivial Z_2 topology. Gapless edge modes are formed at the zone center, where the only band inversion happens. See Fig. 3.10(a). This is the case for Bi_2Se_3, which has an energy "cone." The band inversion (inverted as compared to the atomic limit of orbitals) in Bi_2Se_3 and similar materials is often attributed to the spin-orbit coupling. However, it is not known for sure [46].

In TCI, one needs to pick surface planes that are mirror operation-invariant. This is to ensure that bands on this plane can be classified by their mirror eigenvalues ($\pm i$). One can pick a $k_{x(y,z)}$=0 plane passing through Γ and two L points. See Fig. 3.10(b) (001) surface BZ for reference. Since time-reversal symmetry enforces $C_{+i} = -C_{-i}$, and there are two L points on this 2D Brillouin zone, when the band inversion happens, mirror Chern number $C_M = (C_{+i} - C_{-i})/2$ changes from the trivial $C_M = 0$ to nontrivial by two (an even number). This will lead to C_M Kramers pairs of gapless modes on a mirror-invariant line; see Fig. 3.10(c). For the (111) surface (see Fig. 3.10(b) (111) surface BZ for reference), however, the Dirac cone structure will look like Fig. 3.10(a), and the cone center will be located at the $\bar{L}$, L projected onto the (111) plane.

The structural dynamics study I present in this thesis can couple with the change of topological properties, as structural phase transitions are often accompanied by a change in symmetry. The concepts introduced above can serve as a guide for identifying photoinduced topologically nontrivial phases. I use this procedure to find that Immm phase of SnSe is topologically trivial, in Chap. 5.

References

1. J.P. Heremans, V. Jovovic, E.S. Toberer, A. Saramat, K. Kurosaki, A. Charoenphakdee, S. Yamanaka, G. Jeffrey Snyder, Enhancement of thermoelectric efficiency in PbTe by distortion of the electronic density of states. Science **321**(5888), 554–557 (2008)
2. Y. Pei, A. LaLonde, S. Iwanaga, G. Jeffrey Snyder, High thermoelectric figure of merit in heavy hole dominated PbTe. Energy Environ. Sci. **4**(6), 2085–2089 (2011)
3. L.-D. Zhao, S.-H. Lo, Y. Zhang, H. Sun, G. Tan, C. Uher, C. Wolverton, V.P. Dravid, M.G. Kanatzidis, Ultralow thermal conductivity and high thermoelectric figure of merit in SnSe crystals. Nature **508**(7496), 373–377 (2014)
4. L.-D. Zhao, G. Tan, S. Hao, J. He, Y. Pei, H. Chi, H. Wang, S. Gong, H. Xu, V.P. Dravid, C. Uher, G. Jeffrey Snyder, C. Wolverton, M.G. Kanatzidis, Ultrahigh power factor and thermoelectric performance in hole-doped single-crystal SnSe. Science **351**(6269), 141–144 (2016)
5. C. Chang, M. Wu, D. He, Y. Pei, C.-F. Wu, X. Wu, H. Yu, F. Zhu, K. Wang, Y. Chen, L. Huang, J.-F. Li, J. He, L.-D. Zhao, 3d charge and 2d phonon transports leading to high out-of-plane ZT in n-type SnSe crystals. Science **360**(6390), 778–783 (2018)
6. Z. Zheng, X. Su, R. Deng, C. Stoumpos, H. Xie, W. Liu, Y. Yan, S. Hao, C. Uher, C. Wolverton, M.G. Kanatzidis, X. Tang, Rhombohedral to cubic conversion of GeTe via MnTe alloying leads to ultralow thermal conductivity, electronic band convergence, and high thermoelectric performance. J. Am. Chem. Soc. **140**(7), 2673–2686 (2018)
7. U. Aseginolaza, R. Bianco, L. Monacelli, L. Paulatto, M. Calandra, F. Mauri, A. Bergara, I. Errea, Phonon collapse and second-order phase transition in thermoelectric SnSe. Phys. Rev. Lett. **122**, 075901 (2019)
8. B. Jiang, Y. Yu, J. Cui, X. Liu, L. Xie, J. Liao, Q. Zhang, Y. Huang, S. Ning, B. Jia, B. Zhu, S. Bai, L. Chen, S.J. Pennycook, J. He, High-entropy-stabilized chalcogenides with high thermoelectric performance. Science **371**(6531), 830–834 (2021)
9. M. Wuttig, V.L. Deringer, X. Gonze, C. Bichara, J.-Y. Raty, Incipient metals: functional materials with a unique bonding mechanism. Adv. Mater. **30**(51), 1803777 (2018)
10. D. Lencer, M. Salinga, B. Grabowski, T. Hickel, J. Neugebauer, M Wuttig, A map for phase-change materials. Nat. Mater. **7**(12), 972–977 (2008)
11. S. Lee, K. Esfarjani, T. Luo, J. Zhou, Z. Tian, G. Chen, Resonant bonding leads to low lattice thermal conductivity. Nat. Commun. **5**, 3525 (2014)
12. E.A. Albanesi, C.M.I. Okoye, C.O. Rodriguez, E.L. Peltzer y Blanca, A.G. Petukhov, Electronic structure, structural properties, and dielectric functions of IV-VI semiconductors: PbSe and PbTe. Phys. Rev. B **61**, 16589–16595 (2000)
13. U.V. Waghmare, N.A. Spaldin, H.C. Kandpal, R. Seshadri, First-principles indicators of metallicity and cation off-centricity in the IV-VI rocksalt chalcogenides of divalent Ge, Sn, and Pb. Phys. Rev. B **67**(12), 125111 (2003)
14. C.W. Li, J. Hong, A.F. May, D. Bansal, S. Chi, T. Hong, G. Ehlers, O. Delaire, Orbitally driven giant phonon anharmonicity in SnSe. Nat. Phys. **11**(12), 1063 (2015)
15. G. Lucovsky, R.M. White, Effects of resonance bonding on the properties of crystalline and amorphous semiconductors. Phys. Rev. B **8**(2), 660 (1973)
16. P.B. Littlewood, The crystal structure of IV-VI compounds. I. Classification and description. J. Phys. C: Solid State Phys. **13**(26), 4855–4873 (1980)
17. K. Behnia, Finding merit in dividing neighbors. Science **351**(6269), 124–124 (2016)
18. M.H. Cohen, L.M. Falicov, S. Golin, Crystal chemistry and band structures of the group V semimetals and the IV-VI semiconductors. IBM J. Res. Dev. **8**(3), 215–227 (1964)
19. T. Chattopadhyay, J. Pannetier, H.G. Von Schnering, Neutron diffraction study of the structural phase transition in SnS and SnSe. J. Phys. Chem. Solids **47**(9), 879–885 (1986)
20. S.K. Mishra, S. Satpathy, O. Jepsen, Electronic structure and thermoelectric properties of bismuth telluride and bismuth selenide. J. Phys.: Condensed Matter **9**(2), 461 (1997)
21. L.J. Sham, Electronic contribution to lattice dynamics in insulating crystals. Phys. Rev. **188**, 1431–1439 (1969)

22. W. Porod, P. Vogl, Lattice dynamics and phase transitions in IV–VI compounds, in *Physics of Narrow Gap Semiconductors* (Springer, Berlin, 1982), pp. 247–251
23. A. Bussmann-Holder, H. Bilz, R. Vogl, Electronic and dynamical properties of IV–VI compounds, *Dynamical Properties of IV-VI Compounds* (Springer, Berlin, 1983), pp. 51–98
24. P.B. Littlewood, V. Heine, The infrared effective charge in IV-VI compounds. I. A simple one-dimensional model. J. Phys. C: Solid State Phys. **12**(21), 4431 (1979)
25. P. Vogl, Dynamical effective charges in semiconductors: a pseudopotential approach. J. Phys. C: Solid State Phys. **11**(2), 251 (1978)
26. P.B. Littlewood, The infrared effective charge in IV-VI compounds. II. A three dimensional calculation. J. Phys. C: Solid State Phys. **12**(21), 4441 (1979)
27. T. Lanigan-Atkins, S. Yang, J.L. Niedziela, D. Bansal, A.F. May, A.A. Puretzky, J.Y.Y. Lin, D.M. Pajerowski, T. Hong, S. Chi, G. Ehlers, O. Delaire, Extended anharmonic collapse of phonon dispersions in SnS and SnSe. Nat. Commun. **11**(1), 1–9 (2020)
28. I.T. Witting, T.C. Chasapis, F. Ricci, M. Peters, N.A. Heinz, G. Hautier, G. Jeffrey Snyder, The thermoelectric properties of bismuth telluride. Adv. Electron. Mater. **5**(6), 1800904 (2019)
29. S.-Y. Xu, C. Liu, N. Alidoust, M. Neupane, D. Qian, I. Belopolski, J.D. Denlinger, Y.J. Wang, H. Lin, L.A. a Wray, et al., Observation of a topological crystalline insulator phase and topological phase transition in $Pb_{1-x}Sn_xTe$. Nat. Commun. **3**(1), 1–11 (2012)
30. T.H. Hsieh, H. Lin, J. Liu, W. Duan, A. Bansil, L. Fu, Topological crystalline insulators in the SnTe material class. Nat. Commun. **3**(1), 1–7 (2012)
31. P. Dziawa, B.J. Kowalski, K. Dybko, R. Buczko, A. Szczerbakow, M. Szot, E. Łusakowska, T. Balasubramanian, B.M. Wojek, M.H. Berntsen, et al., Topological crystalline insulator states in $pb_{1-x}sn_xse$. Nat. Mater. **11**(12), 1023–1027 (2012)
32. W. Jin, S. Vishwanath, J. Liu, L. Kong, R. Lou, Z. Dai, J.T. Sadowski, X. Liu, H.-H. Lien, A. Chaney, Y. Han, M. Cao, J. Ma, T. Qian, S. Wang, M. Dobrowolska, J. Furdyna, D.A. Muller, K. Pohl, H. Ding, J.I. Dadap, H.G. Xing, R.M. Osgood, Electronic structure of the metastable epitaxial rock-salt SnSe 111 topological crystalline insulator. Phys. Rev. X **7**, 041020 (2017)
33. H. Zhang, C.-X. Liu, X.-L. Qi, X. Dai, Z. Fang, S.-C. Zhang, Topological insulators in Bi_2Se_3, Bi_2Te_3 and Sb_2Te_3 with a single dirac cone on the surface. Nat. Phys. **5**(6), 438–442 (2009)
34. Y. Ando, Topological insulator materials. J. Phys. Soc. Jpn. **82**(10), 102001 (2013)
35. L. Fu, C.L. Kane, Time reversal polarization and a Z2 adiabatic spin pump. Phys. Rev. B **74**(19), 195312 (2006)
36. R. Resta, Macroscopic polarization in crystalline dielectrics: the geometric phase approach. Rev. Mod. Phys. **66**(3), 899 (1994)
37. E.I. Blount, Formalisms of band theory. J. Phys C: Solid State Phys. **13**, 305–373 (1962)
38. D. Bercioux, J. Cayssol, M.G. Vergniory, M.R. Calvo, *Topological Matter: Lectures from the Topological Matter School 2017*, vol. 190 (Springer, Cham, 2018)
39. L. Fu, C.L. Kane, Ctopological insulators with inversion symmetry. Phys. Rev. B **76**, 045302 (2007)
40. M.S. Dresselhaus, G. Dresselhaus, A. Jorio, *Group Theory: Application to the Physics of Condensed Matter* (Springer Science & Business Media, 2007)
41. J.O. Dimmock, G.B. Wright, Band edge structure of PbS, PbSe, and PbTe. Phys. Rev. **135**, A821–A830 (1964)
42. J.O. Dimmock, I. Melngailis, A.J. Strauss, Band structure and laser action in pb x sn 1- x te. Phys. Rev. Lett. **16**(26), 1193 (1966)
43. D.L. Mitchell, R.F. Wallis, Theoretical energy-band parameters for the lead salts. Phys. Rev. **151**, 581–595 (1966)
44. Y. Ando, L. Fu, Topological crystalline insulators and topological superconductors: from concepts to materials. Annu. Rev. Condens. Matter Phys. **6**(1), 361–381 (2015)
45. J. Liu, W. Duan, L. Fu, Two types of surface states in topological crystalline insulators. Phys. Rev. B **88**(24), 241303 (2013)
46. S. Chege, P. Ningi, J. Sifuna, G.O. Amolo, Origin of band inversion in topological Bi_2Se_3. AIP Adv. **10**(9), 095018 (2020)

Chapter 4
Ultrafast Lasers and X-Ray Pump-Probe Experiment

Abstract This chapter highlights the instrument-intensive nature of ultrafast spectroscopy and the importance of advanced technologies. It provides an overview of the instruments used in optical pump X-ray probe experiments, including the pump laser (near-infrared) and the probe laser (hard X-ray FEL). The chapter discusses both lasers before presenting the integrated setup.

4.1 Ultrafast Near-Infrared Lasers

In 2018, the Nobel Prize in Physics was awarded for the technique of chirped pulse amplification (CPA). The technique allows producing short laser pulses with high peak power and opened doors for laser machining and laser surgery applications. It also facilitates research areas including multiphoton physics (or high harmonic generation, where CPA is essential), nonlinear optics, and pump-probe spectroscopy with sub-ps time resolution. CPA is widely used in modern ultrafast optical lasers, and the most common type runs on Ti:sapphire (titanium doped, the gain comes from Ti^{3+}) as the gain medium. Before the invention of CPA, the amplification of short light pulses was limited by nonlinear effects (such as self-focusing) that cause gain medium damage. The way around is to stretch (chirp) the mode-locked pulse first so that the peak power is low, send the pulse through the gain medium for amplification, and compress the amplified pulse afterward. This idea is illustrated in Fig. 4.1 from the original paper by Strickland and Mourou. Strickland used Nd:YAG laser at her times for the input mode-locked pulses. Today, we mostly see Ti:sapphire lasers as the initial input. She used a 1.4 km communication fiber as a stretcher, as shown in Fig. 4.1. Nowadays, it is more standard to use gratings for both stretcher and compressor, which avoids self-phase modulation in fibers. The regenerative amplifier gain medium in Fig. 4.1, the Nd:glass, is often replaced by Ti:sapphire nowadays, and the Pockels cell can be changed to the acousto-optical modulators, depending on the desired repetition rate of the laser.

Below, I will briefly introduce the key components of an ultrafast optical laser: the oscillator that generates the mode-locked pulse, stretcher and compressor gratings, and the (regenerative) amplifier. The main advantage of Ti:sapphire laser

Y. Huang, *Towards the Optical Control of Resonantly Bonded Materials*, Springer Theses, https://doi.org/10.1007/978-3-031-42826-5_4

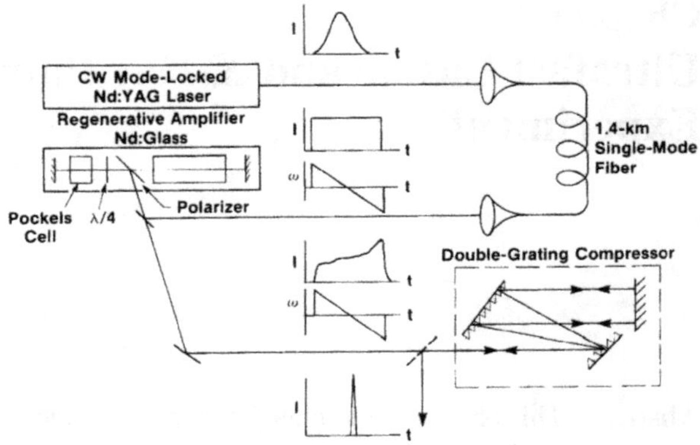

Fig. 4.1 CPA illustration. Figure is reprinted from the original paper [1] that won the 2018 Nobel Prize in Physics, with the permission from Elsevier

is the large gain bandwidth of Ti^{3+} ion, much larger than that of rare earth-doped laser gain media [2], which allows the generation of very short pulses as well as tunability over a wide wavelength window. The Ti^{3+} shifts the atomic position upon excitation, which in turn changes the energy levels, making a four-level laser system, as shown in Fig. 4.2(a) A, B, C, D. The emission terminates on high vibrational levels of the ground state, which is unpopulated because of a fast relaxation rate, favoring population inversion.

Ti:sapphire oscillator produces short pulse by Kerr lens mode-locking. The laser runs in continuous-wave mode after being turned on. Disturbances produced in the cavity can initiate mode-locking (e.g., by shaking the resonant mirror). In essence, Ti:sapphire works both as a saturable absorber and as a Kerr lens. In each resonator round trip, the saturable absorber will then favor the light which has somewhat higher intensities because this light can saturate the absorption slightly more than light with lower intensities. The high intensity then leads to the Kerr lensing effect, a χ^3 effect that changes the refractive index at high laser intensity: the stronger the laser, the tighter the laser beam focus. Prisms placed in the cavity can be used to adjust the phase difference between wavelengths in the desired color window (for Ti:sapphire, it is typically ± 20 nm around 780 nm) so that all colors have the same phase to produce a transform-limited mode-locked pulse. By compensating the phase difference between colors, adjusting the pump power, and moving the cavity mirror (remember that the Kerr lensing effect gives rise to different focal lengths for different colors and that for certain wavelengths to be amplified, the cavity needs to have the right length), one can eventually optimize the cavity for mode-lock lasing over CW lasing.

The compressor design for CPA is first proposed in a 1969 paper [3] as shown in Fig. 4.2(b), and a stretcher can be similarly designed by inserting lenses. It is not hard to show that the pair grating will introduce a quadratic term in the phase (chirp) $\Delta\phi^{\omega} = \frac{\partial^2\phi}{\partial\omega^2}|_{\omega_0}(\omega - \omega_0)^2$. Stretcher has $\frac{\partial^2\phi}{\partial\omega^2}|_{\omega_0} > 0$. It broadens the pulse in

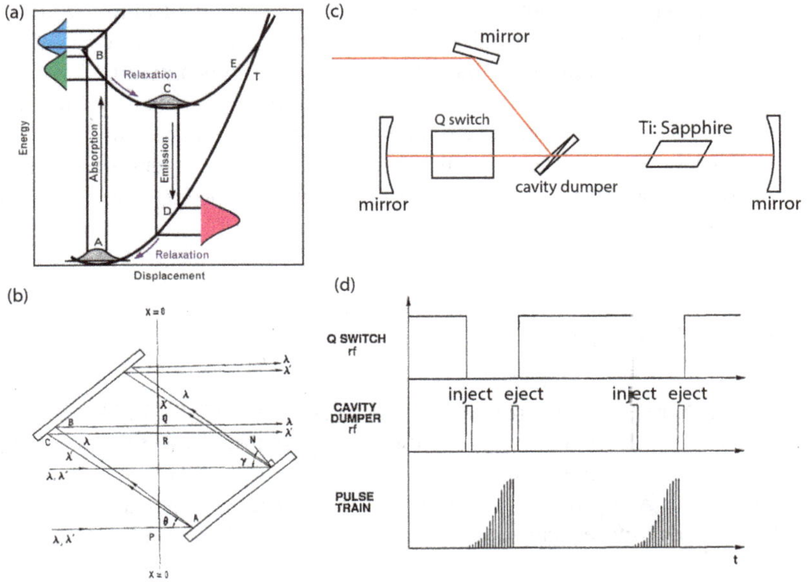

Fig. 4.2 (a) The Ti:sapphire configurational diagram. The energy of the electronic states is plotted as a function of the displacement of the Ti^{3+} ion. Absorption is indicated by the vertical transition from A to B and occurs in the spectrum's blue and green light region. The emission of light is indicated by the transition from C to D and occurs in the red to the infrared region of the spectrum, as indicated at the right. (b) Schematic of a compressor where the red light travels longer than blue light and introduces a negative chirp. λ' is the red light, and λ is the blue light. (c) A crude schematic of Ti:sapphire regenerative amplifier. (d) Time sequence of Q switch, cavity dumper, and the pulse train. Figure (a) is reprinted from ref. [2]. Figure (b) is reprinted from ref. [3], with permission from ©(1969) IEEE. Figure (d) is reprinted from ref. [4] with permission from Optica Publishing Group

the time domain and lowers the peak power.[1] After amplification, the chirped pulse is sent into another pair grating, the compressor, with $\frac{\partial^2 \phi}{\partial \omega^2}|_{\omega_0} < 0$, to remove the chirp. Gratings are linear optical components and do not introduce nonlinear optics effect as the fibers do, and the chirp introduced by a pair grating can, in principle, be entirely removed by another pair grating. The pulse can approach the transform limit if the initial input pulse is transform-limited.

In Fig. 4.2(c), I plot a schematic layout of a regenerative amplifier. The Q switch, when turned on, can break the cavity and prevent lasing to maximize the population inversion in Ti:sapphire, until a seed pulse from the oscillator is injected. The cavity dumper controls the timing of pulse injection and ejection. The seed pulse from the oscillator will be amplified and eventually saturate and feed energy back to the

[1] A stretcher introduces a chirp with a negative sign of group velocity dispersion (GVD). The same sign of GVD produces the rainbows we often see.

medium. Ejection should happen right before the saturation of amplification; see the flattened top of Fig. 4.2(d) pulse train. The Q switch turns on right after the ejection of the amplified pulse so that population inversion is maximized before the next injection. These are conceptual components. In reality, depending on the laser repetition rate, the Q switch and cavity dump can be different types of modulators.

The maximum gain and lasing efficiency for Ti:sapphire lasers are obtained at around 800 nm, and many Ti:sapphire lasers operate with emission wavelengths between about 700 and 900 nm.

4.2 Free-Electron Lasers

The first hard X-ray free-electron lasing was in 2009, at LCLS [5]. The idea of an X-ray free-electron laser (FEL) has been long proposed [6]; however, its feasibility received a lot of questioning until the moment it lased. In Fig. 4.3(a), I show the brightness of the X-ray radiation sources as the instruments get developed since 1990. The famous Moore's law for the chip industry is left far behind. The brightness is defined as

$$B = \frac{\dot{N}_{ph}}{4\pi^2 \sigma_x \sigma_y \Omega \frac{d\omega}{\omega}} \tag{4.1}$$

where the $\dot{N}_{ph}$ is the number of incident photons per second, σ_x, σ_y are the root mean square transverse beam dimensions, Ω is the solid angle spanned by the beam, and $\frac{d\omega}{\omega}$ is the bandwidth in the percentage of the X-ray photon energy.

When transitioning from a tube source to a synchrotron source, the medium changes from solids to relativistic electrons, which removes the constraints of the damage threshold typically associated with solid materials.

The working principle of a synchrotron radiation source involves the passage of electrons through a wiggler (also referred to as an undulator). The wiggler generates a spatially periodic magnetic field, as depicted in Fig. 4.3(b), causing the electrons to follow a wiggling trajectory. As the electrons undergo this bending motion, they experience acceleration and emit radiation.

To estimate the wavelength of the emitted radiation, we consider the following: First, we transform to the electron frame, where the electrons perceive the undulator with a spatial periodicity of L/γ (where $\gamma = \frac{1}{\sqrt{1-\frac{v^2}{c^2}}}$ is the relativistic boost factor, v is the longitudinal speed of the electron, and c is the speed of light). Thus, the wavelength of the radiation emitted, as seen from the electron frame, would be L/γ. However, we need to transform back to the laboratory frame, accounting for a correction factor of $1/2\gamma$ due to the longitudinal Doppler effect caused by the fact that electrons travel in the same direction as the emitted light. This rough estimation tells us the emitted wavelength in the lab frame will be

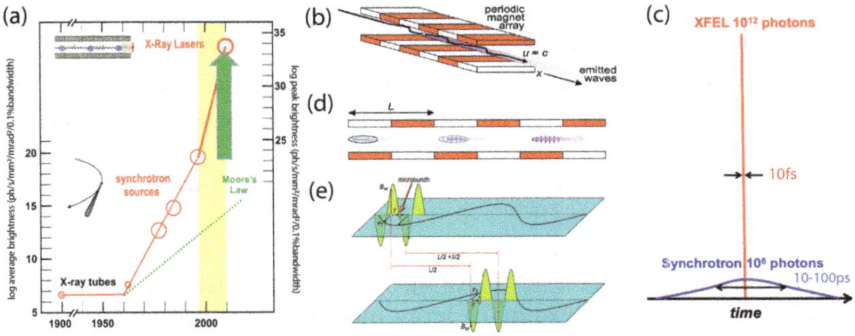

Fig. 4.3 (a) Brightness of X-ray sources improved over the years. (b) Working principles of both synchrotron source and FEL; wiggling electrons radiate X-rays. (c) Comparison of the time structure and the brightness between a FEL X-ray pulse and a synchrotron X-ray pulse. (d) Microbunching process which yields the extremely high brightness of FEL. (e) Resonant condition sustains microbunching and transfer of energy from electron to radiation. Top: radiated E-field is the same direction as the electron transverse velocity v_T. Bottom: after the electron travels over one-half undulator period $L/2$, its transverse velocity v_T is reversed. The wave travels ahead of the electron by one-half wavelength $\lambda/2$; its E-field is also reversed. Figures (a, c) courtesy Zhirong Huang, and figures (b, d, e) are reproduced from ref. [7] with permission from IUCr Journals

$$\lambda = L/2\gamma^2. \tag{4.2}$$

To generate X-rays, especially in the hard X-ray range with wavelengths on the order of angstroms, highly energetic electron beams are required due to the macroscopic scale of the undulator. It is important to consider that the electrons do not travel in a straight path through the undulator, but instead oscillate back and forth under the influence of the magnetic field. This oscillatory motion introduces a correction to the wavelength expression.

The corrected wavelength can be expressed as

$$\lambda = \frac{L}{2\gamma^2}\left(1 + \frac{K^2}{2}\right), \tag{4.3}$$

where $K = \frac{eB_0L}{2\pi mc}$ represents the strength of the magnetic field in the undulator, with e being the elementary charge, B_0 the magnetic field strength, L the undulator periodicity, m the electron mass, and c the speed of light. This equation demonstrates that the X-ray wavelength can be tuned by adjusting the magnetic field strength of the undulator.

The relativistic boost factor γ has additional effects. It enhances the emitted power by a factor of γ^2 and results in a small angular divergence on the scale of $1/\gamma$, both contributing to the increased brilliance (B) of the X-ray source.

Moreover, advancements in electron beam controls allow for the manipulation of electron bunch sizes σ_x and σ_y, enabling them to approach the diffraction limit

$\sigma_x \sigma_y \Omega \rightarrow \lambda^2/4\pi$. This reduction in electron bunch size further enhances the brilliance of the X-ray source.

X-ray FEL provides the brightness that is even many orders of magnitude higher than the synchrotron source. For the same X-ray photon energy, FEL source has 6 orders of magnitude more photons per pulse than the synchrotron source, as shown in Fig. 4.3(c), and in terms of time structure, the pulse width is 3–4 orders of magnitude smaller.

Unlike a typical optical laser that consists of a gain medium, an external pump, and a cavity, a free-electron laser (FEL) operates using a different principle. In an FEL, the gain medium and external energy input are provided by free electrons. However, the absence of hard X-ray mirrors poses a challenge in constructing a cavity based on nominal back reflection, as is the case with optical lasers.

As a result, most hard X-ray FELs operate in what is known as the high-gain single-pass mode. In this mode, electrons travel along a long path before being discarded, without the use of a traditional cavity. This configuration allows for the efficient generation of X-rays without the need for reflective elements commonly found in optical laser cavities. The FEL process utilizes the interaction between the electron beam and an undulator, which produces the necessary amplification and coherence of the X-ray radiation.

The fact that an FEL has a much higher brightness than synchrotron sources has to do with the microbunching of electron bunches. Generally, in a synchrotron source, the electrons in an undulator emit electromagnetic waves without phase correlation. The total emitted intensity will be the sum of intensities produced by individual electrons and is proportional to the number of electrons in the bunch N. In an FEL, the electrons within a microbunch emit with correlated phase—they always wiggle synchronously. It is thus the fields of their emissions that are summed together rather than their intensities. Approximately the radiation intensity is proportional to N^2. We gain by a factor of the number of electrons in a microbunch! See Fig. 4.3(d) for an illustration of the microbunching process as the bunch propagates down the undulator.

Microbunching develops for nearly monoenergetic electrons satisfying the resonant condition Eq. 4.3. Motion of the electrons is determined by their interaction with the combined undulator field and radiation field and can be described by "pendulum equations" in the phase space [8]. Whether electrons gain energy or lose energy from the radiation will depend on the position of the electron relative to the traveling light wave. As the electrons travel, there will be energy modulation and hence the density modulation, of electrons, forming microbunches. The difficulty of producing microbunches for hard X-ray emission is that the higher the photon energy, the larger the γ relativistic factor is required, and the heavier the relativistic longitudinal masses $\propto \gamma^3$, leading to low efficiency in microbunching. Meanwhile, the small microbunch separation at hard X-ray wavelength (λ) sets a high demand for quality beam control.

When the resonant condition is met in a FEL, the transverse velocity component, denoted as v_T, of the electrons maintains its direction relative to the radiated electric field for a significant number of undulator wavelengths. In the laboratory frame, the trajectory of the electrons has a periodicity L that is much larger than the wavelength of the emitted X-rays, denoted as λ (where $\lambda \ll L$).

To understand the process, imagine an electron within a microbunch traveling along the undulator until it reaches the next L where the transverse velocity v_T reverses its direction. At the same time, the velocity difference between the electron and the speed of light $(c - v)$ causes a phase shift between the microbunched electrons and the electromagnetic (EM) wave. Consequently, the radiated electric field, denoted as E, that exerts a force on the same microbunched electrons also reverses its direction. As both v_T and E flip sign, the product $v_T \cdot E$ remains positive see Fig. 4.3(e).

The energy transfer from the electrons to the EM waves occurs due to the work done by the previously emitted electric field on the transverse motion of the electrons. The sign of this work depends on whether energy is being transferred to or from the EM waves. Since the term $v_T \cdot E$ remains positive, indicating a consistent sign, the energy continues to be transferred from the electron beam to the FEL radiation.

FEL has an exponential gain even when operated without a cavity. In ref. [7], it is approximated that the fraction of electrons microbunched (thus emitting in phase) is proportional to $v_T B$. The in-phase radiation, or the energy transferred from the microbunched electrons to the radiation, is thus proportional to $v_T^2 BE \propto I$ where I is the intensity of the EM radiation. Thus, $\frac{dI}{dt} \propto I$, which exactly describes an exponential growth. One can use gain length L_G to characterize the exponential growth $I \propto e^{z/L_G}$. Like usual lasers, FELs saturate in gain when the energy is transferred from the EM field to the gain medium. As the electrons lose energy and their longitudinal velocity gets reduced, the resonant condition Eq. 4.3 is no longer satisfied. At some point in space along the electrons' travel path, the electromagnetic wave will do positive work on and transfer energy back to the electrons; meanwhile, the microbunches are smeared out, leading to gain saturation. $v_T \propto \cos(\frac{2\pi v_L t}{L})$ and $E \propto \cos[2\pi(\frac{(v_L - c)t}{\lambda})]$. The E field reverses sign when $\frac{2\pi \Delta v_L t}{\lambda} = \pi$, and the energy transfer will also reverse. Δv_L is the reduction of longitudinal velocity due to the loss of electron energy. One would want to optimize energy extraction from the electron beam before the energy transfer is reversed, which corresponds to a maximal desirable L_G that at $t = L_G/v_L$, we have $\Delta v_L t = \lambda/2$. A short gain length is desired to maximally transfer electron energy to X-ray photons.

Apart from the high brightness, I emphasize two additional key properties of X-ray FELs. The first is the short pulse duration on the order of 10 fs. The duration of the pulse is determined by the input electron pulse length and can be quite short with a low-charge mode of operation. Although the average X-ray power produced by Linac coherent light source (LCLS), the FEL at SLAC, is quite modest (1 W), the peak power is extremely high (up to 40 GW). The second is the spectrum of X-ray radiation. Self-amplified spontaneous emission (SASE) is the simplest operation of FEL, a process that amplifies radiation from the electron shot noise in a long undulator to produce quasicoherent X-rays. Under SASE mode, the pulse shape changes from pulse to pulse, depending on how each electron bunch initializes the lasing process as it flies and enters the undulator. Each bunch containing several microbunches will produce a spectrum with coherent spikes and a relatively broad bandwidth. An alternative to SASE operation is seeding, for example, using the (hard) X-ray emission from the first half of the undulator to seed the second half [9].

In the previous chapter, I mentioned that diffuse scattering signal is orders of magnitudes weaker than diffraction; thus, we need many photons. I am interested in probing non-equilibrium dynamics on the ultrafast timescale and need a short pulse. FELs can simultaneously meet these demands. Synchrotron source has slicing sources operation [10], but it does worse than FELs in both aspects. During the time of my Ph.D., the hard X-ray still runs a 120 Hz. However, the upgraded superconducting Linac can provide a much higher duty cycle for electron beam production up to MHz range, meaning we will have more photons given a fixed amount of beamtime.

4.3 The Integrated Setup

Figure 4.4 shows both X-ray and optical lasers incident on the sample. All the X-ray scattering experiments I present in later chapters were performed at the X-ray pump-probe (XPP) end-station at LCLS FEL, using 9.5 keV X-rays [11, 12]. The X-rays were monochromatized (to 0.5 eV) using a diamond (111) double

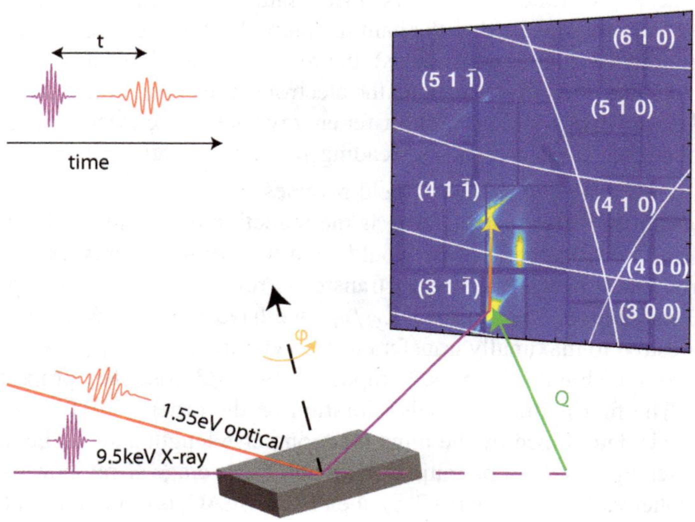

Fig. 4.4 The integrated setup containing the optical laser and the FEL. The sample can be rotated around its normal by an azimuthal angle, and a 2D detector captures the diffuse scattering intensity as a function of delay t between the optical pump and X-ray probe. The green vector shows a scattering wave vector (momentum transfer) **Q**, associated with scattering on a particular pixel on the detector. The detector image shows the typical intensity pattern for a fixed azimuth φ, without pump. White lines represent the mapped-out Brillouin zone boundaries

crystal monochromator (diamond is chosen for its high thermal conductivity), providing nominally $>10^9$ photons per pulse. The near-infrared (NIR) pump pulses with photon energy 1.55 eV were derived from a Ti:sapphire laser. The X-rays illuminated the sample at a nominal grazing incidence of 0.5° with respect to the sample surface to approximately match the penetration depth with the NIR laser, while the NIR beam was nearly colinear at an nominal incident angle of 1°, with p polarization. A fast scan delay stage controlled the nominal delay between the NIR and X-ray pulses. Scattered X-rays were collected by the Cornell-SLAC pixel array detector (CSPAD) [13]. The typical total time resolution is <50 fs, considering the duration of both pulses after correcting for their relative timing jitter. A two-dimensional detector is placed ~100 mm behind the sample, which allows us to map out a large portion of the Ewald sphere. Different regions of the Brillouin zone were accessed by rotating the sample about the sample surface normal (azimuth φ), at nominally fixed grazing incidence. In Fig. 4.4, the detector image is X-ray scattering of a SnSe sample, which I will talk about in Chaps. 5 and 6.

An optical pump X-ray probe experiment generally requires X-ray and optical optics alignment, shot-to-shot spatial and temporal pump-probe overlap, and spectral and spatial beam diagnostics. The data is acquired and stored on a shot-to-shot basis and can easily amount to TBs/day. I will mention two technical aspects of a pump-probe experiment at an FEL from a user's perspective. First of all, getting two beams to overlap in time (on the fs precision) and space (X-ray beam size is on the order of 100 μm) is not trivial. Second, though in principle a 10 fs X-ray pulse and a 10–40 fs optical pulse allow investigations of few-femtosecond dynamics, in reality, there is only limited synchronization between optical laser and FEL, with 100–200 fs r.m.s. jitter limitations. These two aspects, among others, often determine whether you see time-resolved signals or not. In principle, these two problems can be solved by building diagnostic tools based on another pump-probe experiment, with an X-ray pump that generates ultrafast free carriers in a transparent material and an optical probe. The transmission or the transmitted spectrum through the material can be measured. See Fig. 4.5.

At LCLS, achieving the spatial and temporal overlap of two beams at the desired interaction point is accomplished by measuring the transmission of the optical pulse through Si_3N_4 [15]. When the beams overlap, there is a noticeable decrease in transmission, as shown in the right panel of Fig. 4.5. The image of the optical beam, intentionally larger than the X-ray beam, experiences a reduction in intensity at the center where the X-rays interact.

To address timing jitter, a "measure-and-sort" approach is employed [14]. This approach does not eliminate the jitter but allows for its correction during off-fly analysis. In the left panel of Fig. 4.5, the spectrally encoded timing tool used at LCLS is illustrated. A chirped white light with a characterized spectral wavelength-to-time mapping is sent through a transparent YAG material (often used instead of Si_3N_4 as in [14]). The X-ray pulse generates carriers that reduce the transmission for the portion of the spectrum that arrives after the X-ray pulse, resulting in an edge in the transmitted spectrum. The position of this edge can be precisely determined (around 5 fs), enabling the correction of timing jitter.

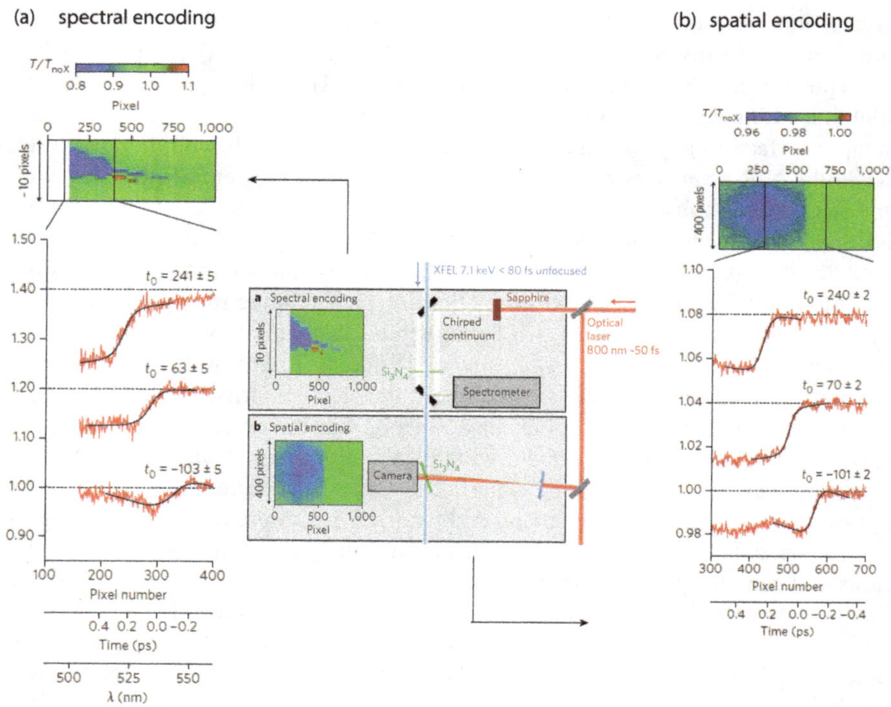

Fig. 4.5 (a) The relative delay between laser and X-ray is encoded into wavelength by using broadband chirped supercontinuum. The camera is positioned after the spectrometer to record the spectrum. The panel on the left shows three spectra; they are line-outs from one-dimensional camera projections for three different shots at the same nominal delay. The edge position of optical laser light transmission encodes the relative delay between pulses, due to the time-wavelength correlation of a chirped pulse, and moves from shot to shot due to the timing jitter between the optical and X-ray laser pulses. (b) The temporal delay between laser and X-ray is spatially encoded. The wavefront of X-ray and optical laser cross at a large angle and different portion of the optical wavefront (transverse to the beam propagating direction) intersects X-ray at different times, leaving an optical transmission edge in the camera that encodes the relative delay between pulses. The right panel shows the camera image and the transmission profiles each encoding a delay time with an edge. All signal images are background (X-ray off) subtracted. Figures are reproduced from ref. [14] with permission from Springer Nature

Alternatively, timing correction can be achieved through spatial encoding. In this method, the white light is crossed with the X-ray beam at a large angle on a Si_3N_4 sample, and the edge separating the shadowed and illuminated areas on the Si_3N_4 camera encodes the relative timing between the two pulses. This technique is employed at SACLA, the FEL in Japan [16].

Future FELs may employ different timing schemes to actively synchronize the FEL and the optical laser. Additionally, the high repetition rate (MHz) upgrade at LCLS holds the potential to benefit the phase stabilization of electronics, potentially rendering timing jitter problems obsolete in the near future.

References

1. D. Strickland, G. Mourou, Compression of amplified chirped optical pulses. Opt. Commun. **55**(6), 447–449 (1985)
2. K. Wall, A. Sanchez, Titanium sapphire lasers. Lincoln Laborat. J. **3**(3), 447–462 (1990)
3. E. Treacy, Optical pulse compression with diffraction gratings. IEEE J. Quantum Electron. **5**(9), 454–458 (1969)
4. T.B. Norris, Femtosecond pulse amplification at 250 khz with a TI: sapphire regenerative amplifier and application to continuum generation. Opt. Lett. **17**(14), 1009–1011 (1992)
5. P. Emma, R. Akre, J. Arthur, R. Bionta, C. Bostedt, J. Bozek, A. Brachmann, P. Bucksbaum, R. Coffee, F.-J. Decker et al., First lasing and operation of an ångstrom-wavelength free-electron laser. Nat. Photon. **4**(9), 641 (2010)
6. John MJ. Madey, Stimulated emission of bremsstrahlung in a periodic magnetic field. J. Appl. Phys. **42**(5), 1906–1913 (1971)
7. G. Margaritondo, P.R. Ribic, A simplified description of X-ray free-electron lasers. J. Synchrot. Rad. **18**(2), 101–108 (2011)
8. Z. Huang, K.J. Kim, Review of X-ray free-electron laser theory. Phys. Rev. Spec. Top. Accelerat. Beams **10**(3), 1–26 (2007)
9. J. Amann, W. Berg, V. Blank, F.-J. Decker, Y. Ding, P. Emma, Y. Feng, J. Frisch, D. Fritz, J. Hastings et al., Demonstration of self-seeding in a hard-x-ray free-electron laser. Nat. Photon. **6**(10), 693–698 (2012)
10. R.W. Schoenlein, S. Chattopadhyay, H.H.W. Chong, T.E. Glover, P.A. Heimann, C.V. Shank, A.A. Zholents, M.S. Zolotorev, Generation of femtosecond pulses of synchrotron radiation. Science **287**(5461), 2237–2240 (2000)
11. M. Chollet, R. Alonso-Mori, M. Cammarata, D. Damiani, J. Defever, J.T. Delor, Y. Feng, J.M. Glownia, J.B. Langton, S. Nelson et al., The X-ray pump–probe instrument at the linac coherent light source. J. Synchrot. Radiat. **22**(3), 503–507 (2015)
12. D. Zhu, A. Robert, T. Henighan, H.T. Lemke, M. Chollet, J.M. Glownia, D.A. Reis, M. Trigo, Phonon spectroscopy with sub-mev resolution by femtosecond X-ray diffuse scattering. Phys. Rev. B **92**, 054303 (2015)
13. P. Hart, S. Boutet, G. Carini, M. Dubrovin, B. Duda, D. Fritz, G. Haller, R. Herbst, S. Herrmann, C. Kenney, N. Kurita, H. Lemke, M. Messerschmidt, M. Nordby, J. Pines, D. Schafer, M. Swift, M. Weaver, G. Williams, D. Zhu, N.V. Bakel, J. Morse, The CSPAD megapixel x-ray camera at LCLS, in *X-Ray Free-Electron Lasers: Beam Diagnostics*, ed. by S.P. Moeller, M. Yabashi, S.P. Hau-Riege. Beamline Instrumentation, and Applications, vol. 8504 (International Society for Optics and Photonics, SPIE, Bellingham, 2012), pp.51–61
14. M. Harmand, R. Coffee, M.R. Bionta, M. Chollet, D. French, D. Zhu, D.M. Fritz, H.T. Lemke, N. Medvedev, B. Ziaja et al., Achieving few-femtosecond time-sorting at hard X-ray free-electron lasers. Nat. Photon. **7**(3), 215–218 (2013)
15. T. Sato, J.M. Glownia, M.R. Ware, M. Chollet, S. Nelson, D. Zhu, A simple instrument to find spatiotemporal overlap of optical/X-ray light at free-electron lasers. J. Synchrot. Radiat. **26**(3), 647–652 (2019)
16. K. Nakajima, Y. Joti, T. Katayama, S. Owada, T. Togashi, T. Abe, T. Kameshima, K. Okada, T. Sugimoto, M. Yamaga, T. Hatsui, M. Yabashi, Software for the data analysis of the arrival-timing monitor at SACLA. J. Synchrot. Radiat. **25**(2), 592–603 (Mar2018)

Chapter 5
Photoinduced Novel Lattice Instability in SnSe

Abstract We present a novel nonthermal lattice instability in SnSe, an IV–VI semiconductor, induced by ultrafast above-gap excitation. This instability is accompanied by a significant softening of the lowest-frequency A_g phonon, previously associated with the thermally driven phase transition to a Cmcm structure. However, by a quantitative reconstruction of the atomic displacements from time-resolved X-ray diffraction, we show that ultrafast photoexcitation with near-infrared (1.55 eV) light induces a distortion toward a different structure with Immm symmetry. The Immm structure of SnSe is an orthorhombic distortion of the rock salt structure and does not occur in equilibrium. These findings underscore the necessity of employing ultrafast structural probes to unveil subtle atomic-scale dynamics that remain elusive or undetectable using conventional spectroscopic techniques.

In this chapter, part of the figures are reprinted from Ref. [1] with permission from the American Physical Society. DFT calculations are provided by collaborators Shan Yang and Olivier Delaire at Duke University.

5.1 Structural Phases of SnSe

Under ambient conditions, SnSe stabilizes in a centrosymmetric layered orthorhombic Pnma structure [2, 3]; see Fig. 5.1. Compared to the rock salt parent structure, the Pnma phase breaks the symmetry between the six nearest neighbor bonds that connect atoms of different elemental species, featuring alternate shearing of the bilayers accompanied by the lengthening and rotation of d_1 and d_4, as well as buckling and off-centering in the $\mathbf{b} - \mathbf{c}$ plane network formed by d_2 and d_3. SnSe undergoes a second-order phase transition [5] at high temperature (807 K) [5, 6] or pressure (10.5 GPa) [7] to a higher symmetry orthorhombic phase with space group Cmcm, where further shearing of the bilayers causes the Sn and Se atoms to align in the $\mathbf{a} - \mathbf{b}$ plane as indicated with the dashed orange line in Fig. 5.1. In this work, we find that photoexcitation induces an instability toward a new structure

Y. Huang, *Towards the Optical Control of Resonantly Bonded Materials*, Springer Theses, https://doi.org/10.1007/978-3-031-42826-5_5

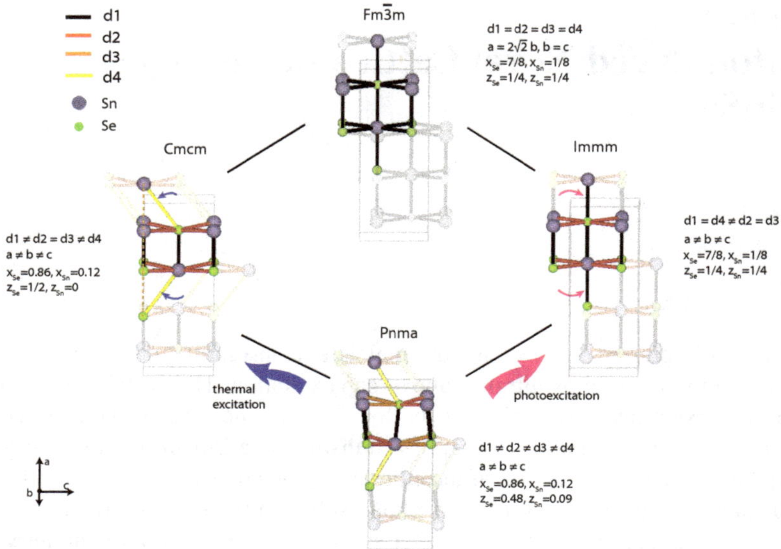

Fig. 5.1 Relations between local coordination and atomic positions for different SnSe structures. At ambient conditions, the Pnma structure has the Sn and Se atoms off-center in the **b** − **c** plane and is heavily distorted from the symmetric parent cubic structure Fm$\bar{3}$m ($d_1 = d_4 = d_2 = d_3$). Bonds of the same color are equivalent under the symmetry of the given lattice. Above 807 K, SnSe stabilizes in Cmcm ($d_1 \neq d_2 = d_3 \neq d_4$) where d_4 bonds rotate further away from the parent rock salt structure compared to Pnma. Orange broken lines in the Cmcm structure highlight the atoms located in the same **a** − **b** plane. Under photoexcitation, the atoms move toward the Immm structure, which is the highest symmetry orthorhombic distortion of the rock salt structure ($d_1 = d_4 \neq d_2 = d_3$). We parameterize all crystal structures using the orthorhombic Pnma primitive unit cell as shown here, unless otherwise stated. The atomic positions are specified by $\pm(x_s, \frac{1}{4}, z_s)$ and $\pm(\bar{x}_s + \frac{1}{2}, \frac{3}{4}, z_s + \frac{1}{2})(s \in \{\text{Sn, Se}\})$, where x_s and z_s are the fractional coordinates along **a** and **c**-axis. Atoms in both the Immm and Fm$\bar{3}$m structures are located at high symmetry positions $x_{\text{Se}} = \frac{7}{8}, z_{\text{Se}} = \frac{1}{4}, x_{\text{Sn}} = \frac{1}{8}, z_{\text{Sn}} = \frac{1}{4}$. In the Cmcm phase, atoms occupy different high symmetry positions $z_{\text{Sn}} = 0, z_{\text{Se}} = \frac{1}{2}$. The Pnma fractional coordinates are taken from DFT calculations, and the Cmcm fractional coordinates are taken from Ref. [2]. Visualization made with VESTA [4]

distinct from this high-temperature Cmcm phase. As our analysis below shows, this lattice instability is toward an Immm structure, the highest symmetry orthorhombic distortion of rock salt, where there is no off-centering, and the bilayers are not sheared ($d_1 = d_4$ and $d_2 = d_3$ as shown in Fig. 5.1). It is crucial to note that the Cmcm and Immm structures do not possess a supergroup-subgroup relationship. Consequently, it is not possible to transform one structure into the other through a displacement that preserves all of their respective symmetries (A_g modes).

The space groups discussed in this chapter (Pnma, Cmcm, Pmmn, Immm) are subgroup descendants of Fm$\bar{3}$m. Their relations are summarized in Fig. 5.2. All the discussed structures are parametrized with Pnma 4c Wyckoff sites [9] ($x_{\text{Sn}}, z_{\text{Sn}}, x_{\text{Se}}, z_{\text{Se}}$) in Table 5.1. Atoms are located at $\pm(x_s, \frac{1}{4}, z_s)$ and $\pm(\bar{x}_s +$

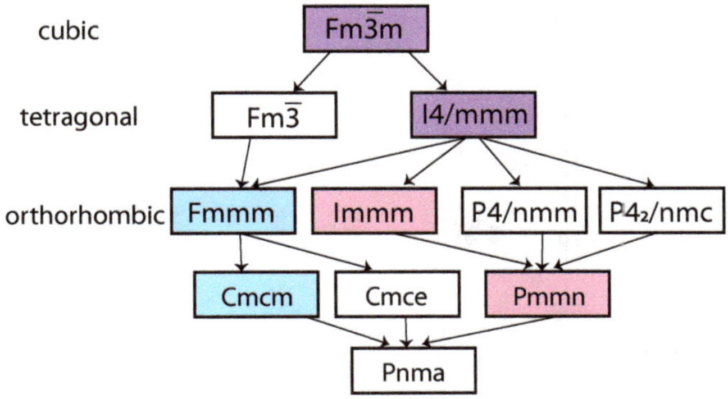

Fig. 5.2 Subgroup descendants of Fm$\bar{3}$m. The descendants of Fm$\bar{3}$m are branched into the Fmmm (blue) and Immm (pink), which separately branches down to Cmcm and Pmmn, which have a common child of Pnma. Cmcm is not a subgroup of Immm. Analysis tools provided by Bilbao Crystallographic Server [8]

Table 5.1 Wyckoff sites of Pnma SnSe and the higher symmetry structures Cmcm, Pmmn, and Immm. x_s and z_s are expressed in fractional coordinates

Pnma 4c	Cmcm 4c	Pmmn 2b	Immm 2d(Se)	Immm 2b(Sn)
x_s,1/4,z_s	x_s,1/4,0	x_s,1/4,1/4	−1/8,−3/4,1/4	−7/8,−3/4,1/4
1/2−x_s,3/4,1/2+z_s	1/2−x_s,3/4,1/2	1/2−x_s,3/4,3/4	−3/8,−1/4,3/4	−5/8,−1/4,3/4
−x_s,3/4,−z_s	−x_s,3/4,0	−x_s,3/4,3/4	−7/8,−1/4,3/4	−1/8,−1/4,3/4
1/2+x_s,1/4,1/2−z_s	1/2+x_s,1/4,1/2	1/2+x_s,1/4,1/4	−5/8,−3/4,1/4	−3/8,−3/4,1/4

$\frac{1}{2}, \frac{3}{4}, z_s + \frac{1}{2}$)($s \in$ {Sn, Se}). Structures in which atoms occupy high symmetry positions can result in a reduction of the number of A_g modes and the elimination of certain free parameters. For instance, the Pnma structure exhibits four A_g modes. In the Cmcm structure, two of the four A_g modes become zone boundary modes, and the unit cell is halved. In the Immm structure, all A_g modes are eliminated, and the unit cell is quartered.

The coordinates listed in Table 5.1 are obtained by converting their respective standard Wyckoff positions to a Pnma basis using the following transformation $(x, y, z)^T|_{\text{Pnma}} = P[(x, y, z)^T + v]|_{\text{Cmcm},Immm}$ where for Immm (conventional unit cell)

$$P = \begin{bmatrix} 0 & 0 & \frac{1}{2} \\ 0 & 1 & 0 \\ -1 & 0 & 0 \end{bmatrix} \tag{5.1}$$

and $v = (\frac{1}{4}\frac{1}{4}\frac{1}{4})^T$. For a Cmcm (conventional unit cell) to Pnma transformation,

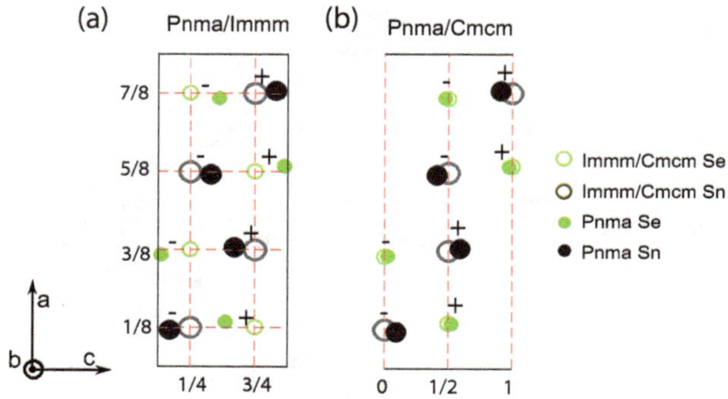

Fig. 5.3 Comparison of the conventional Pnma unit cell with the Pnma/Immm and Pnma/Cmcm structures. Atoms of Pnma phase are marked with filled circles. Atoms in the Immm and Cmcm phase (open circle) are located in different high symmetry mirror planes (represented with dashed red lines). In all the structures, "+" means $y_s = \frac{3}{4}$, and "−" means $y_s = \frac{1}{4}$

$$P = \begin{bmatrix} 0\,1\,0 \\ 0\,0\,1 \\ 1\,0\,0 \end{bmatrix}, \tag{5.2}$$

and $v = (000)^T$ [10]. In this chapter, the fractional coordinates in real space are referenced to the conventional Pnma unit cell, and the wave vectors in reciprocal space are expressed in the Brillouin zone of the Pnma structure, unless stated otherwise. Figure 5.3 depicts the atomic positions of SnSe phases and illustrates the symmetry mirror elements of Cmcm and Immm. It should be noted that atoms occupy different high symmetry positions in Immm and Cmcm. For transformations between Pnma, Cmcm, and Immm Brillouin zones, readers can refer to Fig. 5.4.

5.2 Ultrafast X-Ray Diffraction of SnSe

The experiment was conducted at the X-ray pump-probe (XPP) end-station of the Linac coherent light source (LCLS) X-ray free-electron laser (FEL). A Ti:sapphire laser generated near-infrared (NIR) pump pulses with a photon energy of 1.55 eV, while the X-ray probe pulses had a photon energy of 9.5 keV. The X-rays were monochromatized using a diamond (111) double crystal monochromator, delivering a flux of over 10^9 photons per pulse. The sample used was a single crystal of SnSe grown through a Bridgman-type technique and polished to expose the [100] surface. The X-rays were incident on the sample at a grazing angle of 0.5° to match the penetration depth with the NIR laser, while the NIR beam was nearly colinear at an incident angle of 1°. The temporal delay between the NIR pump and X-ray probe pulses was controlled using a fast scan delay stage. Scattered X-rays were detected

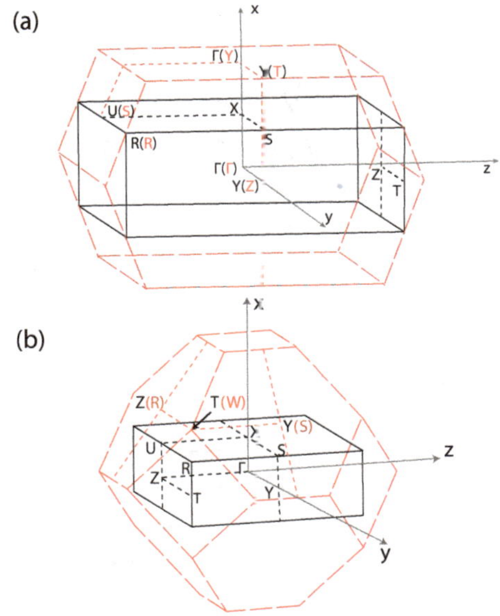

Fig. 5.4 The Brillouin zones of Pnma (black) and in red, Cmcm (a) and Immm (b). The red symbols denote the high symmetry points of the Brillouin zone of Cmcm (a) or Immm (b), and black symbols are those for Pnma. The **x**, **y**, **z** axes overlap with Pnma **a**, **b**, **c** axes, respectively

by the Cornell-SLAC pixel array detector (CSPAD), and both X-ray and NIR pulses had durations of less than 50 fs. The relative arrival time between the X-ray and NIR pulses was determined on a shot-by-shot basis, and the X-ray scattering images were binned with 33 fs intervals based on the sorted arrival times. By rotating the sample about the surface normal (azimuth), multiple X-ray diffraction peaks were accessed while maintaining a fixed grazing incidence angle.

Figure 5.5(a) shows $\tilde{I}_{hkl} = I_{hkl}(t)/I_{hkl}(t < 0)$ where $I_{hkl}(t)$ is the integrated intensity of the Bragg peak for hkl = $(21\bar{1})$, (611), (301), and $(4\bar{1}0)$.[1] The data were taken at a nominal absorbed fluence of $0.8\,\text{mJ/cm}^2$. The traces are offset vertically for clarity. We observe a combination of coherent oscillations in time, the frequency of which is associated with three of the four A_g Raman-active modes of the Pnma structure [11]. Assuming the normalized intensity of the Bragg peaks $\tilde{I}_{hkl}(t > 0)$ is composed of damped harmonic oscillators, we use linear prediction to decompose the time-domain data [12] and obtain robust, highly reproducible oscillator parameters. The normalized intensity of each individual Bragg peak is well described by a sum of decaying cosines such that

$$\tilde{I}_{hkl}(t) = 1 + \sum_i B_{hkl}^{(i)}(1 - e^{-\gamma_i t}\cos(2\pi\nu_i t + \phi_i)), \tag{5.3}$$

[1] $t < 0$ corresponds to the X-ray probe pulse arriving before the NIR pump. We also note that $I(t < 0)$ is virtually identical to the diffraction intensity without the pump.

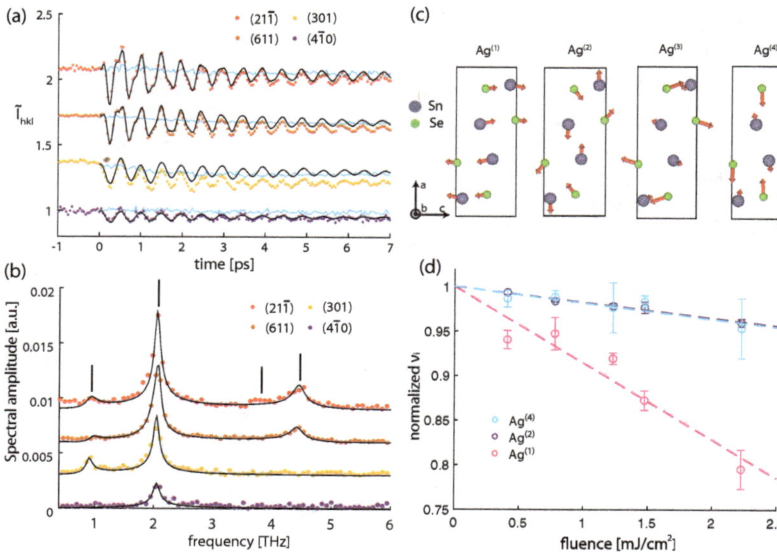

Fig. 5.5 (a) Normalized diffraction intensity $\tilde{I}_{hkl}(t)$ for $hkl = (4\bar{1}0)$, (301), (611), $(21\bar{1})$ Bragg peaks at a nominal absorbed fluence of 0.8 mJ/cm^2. Filled circles: experimental data. Black lines: decomposition of the data in the form of Equation (5.3). Traces are offset vertically for clarity. Blue lines: data subtracted by the most significant components. (b) Fourier transform (colored dots) of the data in (a), and sum of Lorentzians with frequency ν_i, damping γ_i, phase ϕ_i, and amplitude $B_{hkl}^{(i)}$ retrieved from the decomposition (black line). Three out of the four Raman-active A_g modes in Pnma SnSe are observed in the time-resolved X-ray scattering data. The frequencies of the A_g modes measured in Raman spectroscopy are indicated with short black bars. (c) Eigendisplacements ($\times$ 30) of the four A_g modes. (d) Normalized mode frequencies as a function of nominal absorbed fluence from the analysis of the $(21\bar{1})$ Bragg peak

with up to three components i. We absorb the sign of the initial intensity change into $B_{hkl}^{(i)}$ and find ϕ_i to be within $\pm 0.1\pi$. The small absolute value of ϕ_i is consistent with a macroscopic atomic motion produced via displacive excitation of coherent phonons (DECP) induced by above-gap excitation [13] and the ensuing interatomic force changes [14]. DECP typically involves A_g phonons [13, 15] which fully respect the symmetry of the initial state and can potentially connect to higher symmetry phases via displacive phase transitions [13]. The decomposition of the experimental data in the form of Equation (5.3) is shown with black lines in Fig. 5.5(a) for a nominal incident fluence of 0.8 mJ/cm^2. The inclusion of only three oscillators captures well the observed time dependence. The residuals (blue lines in Fig. 5.5(a)) show a slowly varying, non-exponential background. This slowly varying background is likely due to strain that develops and propagates over the probed volume on a much longer timescale than the optical phonons [16]. Thus, the dynamics of the relevant materials can be considered as occurring at a constant volume.

The colored dots in Fig. 5.5(b) show the magnitude of the Fourier transform of the data in Fig. 5.5(a). The black lines in (b) show the sum of Lorentzians obtained from the fitted frequency, amplitude, and damping constant in Equation (5.3).[2] The data clearly reveal three modes at 0.9, 2, and 4.5 THz, in agreement with the frequencies of the $A_g^{(1)}$, $A_g^{(2)}$, and $A_g^{(4)}$ modes observed in Raman measurements [11]. While the $A_g^{(3)}$ mode is observable in ultrafast pump-probe optical reflectivity and Raman scattering measurements, the X-ray data does not exhibit a pronounced signature of the $A_g^{(3)}$ mode above the noise level (as seen in the residual traces in light blue in Fig. 5.5(a)). This is likely attributed to the relatively short lifetime of the $A_g^{(3)}$ mode and the limited sensitivity of the measured Bragg peaks to the $A_g^{(3)}$ motion. The eigendisplacements (×30) of the four A_g modes obtained from harmonic phonon calculation with DFT are shown in Fig. 5.5(c).[3] The $A_g^{(1)}$ displacements primarily involve shearing between bilayers along the **c**-axis, while $A_g^{(2)}$ primarily affects the buckling of the bilayer by displacing Sn and Se atoms from the same atomic layer in opposite directions along the **a**-axis. The $A_g^{(3)}$ and $A_g^{(4)}$ displacements are similar to those of the $A_g^{(1)}$ and $A_g^{(2)}$ modes, respectively, but with Sn atoms π out of phase. It is important to note that in DECP, depending on the initial phase of motion, the direction of atomic motion can be opposite to what is depicted in Fig. 5.5(c).

The $A_g^{(1)}$ phonon mode is of particular interest because it becomes unstable as the temperature approaches the Pnma-Cmcm phase transition at $T_c = 807\,\text{K}$ [5], and it strongly overlaps with the order parameter of this second-order phase transition [5, 6, 19, 20]. At $T > T_c$, this mode becomes an acoustic mode at the zone boundary Y of the Cmcm Brillouin zone. Figure 5.5(d) shows the normalized mode frequency as a function of nominal absorbed fluence. The y-axis represents the normalized fitted frequencies, denoted as ν_i $(i \in 1, 2, 4)$, obtained from Equation (5.3) by extrapolating their values to zero fluence. The $A_g^{(1)}$ mode exhibits a significant softening (decrease in frequency) of up to 20% at the highest absorbed fluence in the experiment (2.2 mJ/cm^2), while the $A_g^{(2)}$ and $A_g^{(4)}$ modes show a less drastic softening of less than 4%. The error bars on the frequencies represent the

[2] The spectral amplitude of $\tilde{I}_{hkl}$ is $\left|\sum_i B_{hkl}^{(i)} \frac{-(2\pi\nu_i)\sin\phi_i+(\gamma_i-2\pi i\nu_i)\cos\phi_i}{(2\pi\nu_i)^2+(\gamma_i-2\pi i\nu_i)^2}\right|$.

[3] From DFT, we obtain interatomic force constants and recast them into the dynamical matrix $D(\mathbf{q})$ [17, 18]. The solutions to $D(\mathbf{q})\boldsymbol{\epsilon}_{\mathbf{q}}^{(i)} = \omega_i^2(\mathbf{q})\boldsymbol{\epsilon}_{\mathbf{q}}^{(i)}$ yield the orthonormal eigenvectors $\boldsymbol{\epsilon}_{\mathbf{q}}^{(i)}$, where $\mathbf{q}$ represents the wave vector. For a SnSe conventional unit cell that contains 8 atoms, the eigenvector $\boldsymbol{\epsilon}_{\mathbf{q}}^{(i)}$ should contain $8 \times 3 = 24$ elements. We only discuss zone center phonon ($\mathbf{q} = \mathbf{0}$) and simplify the notation $\boldsymbol{\epsilon}_{\mathbf{q}=\mathbf{0}}^{(i)}$ to $\boldsymbol{\epsilon}^{(i)}$. The $A_g^{(i)}$ displacement of the s atom $\mathbf{u}_s$

$$\mathbf{u}_s^{(i)} = \text{Re}\left\{\alpha_i \frac{1}{\sqrt{\mu_s}} \boldsymbol{\epsilon}_s^{(i)} \exp[-i\omega_i t]\right\} \tag{5.4}$$

is proportional to the reduced amplitudes α_i, which are scalars, and the eigendisplacement, which is the eigenvector normalized by square root of atomic masses. $\boldsymbol{\epsilon}_s^{(i)}$ is the atomic projection of the eigenvector $\boldsymbol{\epsilon}^{(i)}$ and is a (3×1) vector, as is $\boldsymbol{u}_s^{(i)}$.

statistical errors caused by shot-to-shot fluctuations in the FEL intensity. These errors are estimated using the standard error of the ensemble of ν_i values obtained by decomposing random sub-samples of the data. The pronounced softening of ν_1 indicates the presence of a lattice instability, specifically a softening of the interatomic potential associated with the mode coordinate. However, through a thorough analysis of $\tilde{I}_{hkl}(t)$, we demonstrate that this instability is not related to the thermal transition to Cmcm, but rather toward a structure that exhibits similarities to Immm for large displacements.

In the kinematic diffraction limit, the intensity of the (hkl) Bragg peak using the Pnma unit cell convention is

$$\tilde{I}_{hkl}(t) \propto \Big| \sum_{s \in \text{Se,Sn}} 4 f_s \cos\left[2\pi\left(h x_s(t) + \frac{h+k+l}{4}\right)\right] \times \cos\left[2\pi\left(l z_s(t) - \frac{h+l}{4}\right)\right] \Big|^2, \tag{5.5}$$

where f_s are the atomic form factors. The four parameters $x_s, z_s (s \in \{\text{Sn, Se}\})$ fully represent the atomic positions of the eight atoms in a Pnma unit cell in fractional coordinates $\pm(x_s, \frac{1}{4}, z_s)$ and $\pm(\bar{x}_s + \frac{1}{2}, \frac{3}{4}, z_s + \frac{1}{2})$ [9]. According to Equation (5.5), the intensity of Bragg peaks where $h + l = \text{odd}$ (e.g., $(21\bar{1})$ and (611)) decreases monotonically to zero as $z_{\text{Sn}} \rightarrow 0$ and $z_{\text{Se}} \rightarrow \frac{1}{2}$ in the Cmcm phase. However, in Fig. 5.6(a), we observe the mode decomposition of the $(21\bar{1})$ data, where the

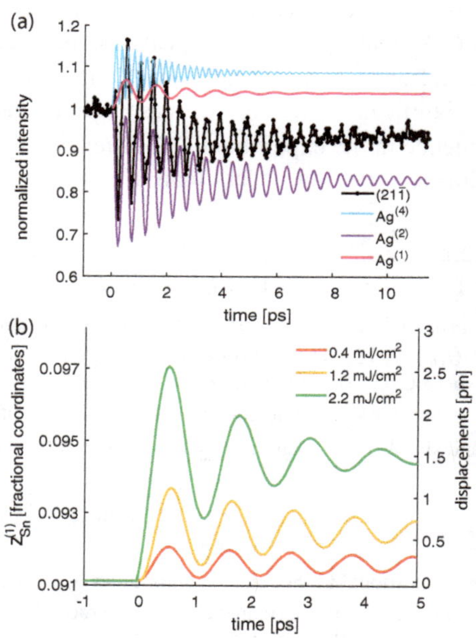

Fig. 5.6 (a) Decomposition of the $(21\bar{1})$ time-resolved diffraction signal at a nominal absorbed fluence of $0.8\,\text{mJ/cm}^2$ according to Equation (5.3). (b) Time dependence of $z^{(1)}_{\text{Sn}}$, the Sn z position projected onto the $A_g^{(1)}$ mode, as calculated from the $(21\bar{1})$ peak measured under different nominal absorbed fluences. $z^{(1)}_{\text{Sn}}$ is defined in fractional coordinates. The right y-axis shows the corresponding displacements in picometers (pm)

$A_g^{(1)}$ component (pink trace) oscillates around an increased intensity ($\tilde{I}_{hkl} > 1$). This behavior suggests that $z_{\mathrm{Sn}}^{(1)}$, the $A_g^{(1)}$ mode-projected z_{Sn} motion, moves *away* from zero rather than toward it. Figure 5.6(b) displays the measurements of $z_{\mathrm{Sn}}^{(1)}$ at different fluences on the $(21\bar{1})$ Bragg peak. Supplementary mode decompositions featuring $A_g^{(1)}$ for other Bragg peaks can be found in the Appendix. Since the $A_g^{(1)}$ mode primarily involves z_{Sn} motion and strongly overlaps with the order parameter of the Pnma-Cmcm phase transition, photoexcitation of the $A_g^{(1)}$ mode appears to displace the lattice further away from the Cmcm structure.

5.3 Quantitative Reconstruction of Photoexcited Atomic Motion

To refine the above observation, we quantitatively reconstruct the photoexcited atomic motion using the diffraction data. We will show below in Fig. 5.7(a) that the observed $\Delta z_{\mathrm{Sn}}^{(1)} > 0$ which we illustrate using Bragg peak $(21\bar{1})$ as an example in Fig. 5.6 is consistent for all measured Bragg peaks and excitation fluences. We first use Equation (5.3) to obtain $B_{hkl}^{(i)}$ for each individual Bragg peak (hkl) and each excitation fluence. Then based on Equation (5.5), we use $B_{hkl}^{(i)}$ and the eigendisplacements, shown in Fig. 5.5(c), to obtain a dimensionless amplitude α_i. Figure 5.6(b) shows the dynamics of $z_{\mathrm{Sn}}^{(1)}$, the $A_g^{(1)}$-projected z_{Sn} displacements, extracted from the $(21\bar{1})$ peak under nominal absorbed fluences 0.4, 1.2, and 2.2 mJ/cm^2. For reference, we also show the $z_{\mathrm{Sn}}^{(1)}$ displacement in picometers (pm) on the right y-axis in Fig. 5.6(b). The $A_g^{(1)}$ motion increases z_{Sn} from the initial value $z_{\mathrm{Sn}} = 0.09$, instead of decreasing it toward $z_{\mathrm{Sn}} = 0$, the atomic position of the Cmcm phase. The fact that photoexcitation both softens the $A_g^{(1)}$ mode and produces a shift in its quasi-equilibrium position *further away* from Cmcm signals an instability associated with a new transient structure.

The transient structure can be understood by combining all A_g displacements and reconstructing the overall lattice distortion in the photoexcited state. To consistently incorporate *all* data sets with a variety of pump fluences and Bragg peaks, we take the degree of $A_g^{(2)}$ frequency softening as a more accurate measure of the absorbed fluence than the nominal absorbed fluence, which is obtained from the measured pulse energy and the illuminated area.[4] The reduced mode amplitudes α_i correlate linearly with $A_g^{(2)}$ mode frequencies as the absorbed fluence varies, as shown in Fig. 5.7(a). The dashed lines in Fig. 5.7(a) are fits of $\alpha_i = b_i(\nu_2 - \nu_2^0)$, where b_i

[4] This is to mitigate the effect of systematic errors due to the anisotropy of refractive index, which results in variations of optical absorption caused by azimuthal rotations. $A_g^{(2)}$ has a good signal-noise ratio in all the measured Bragg peaks.

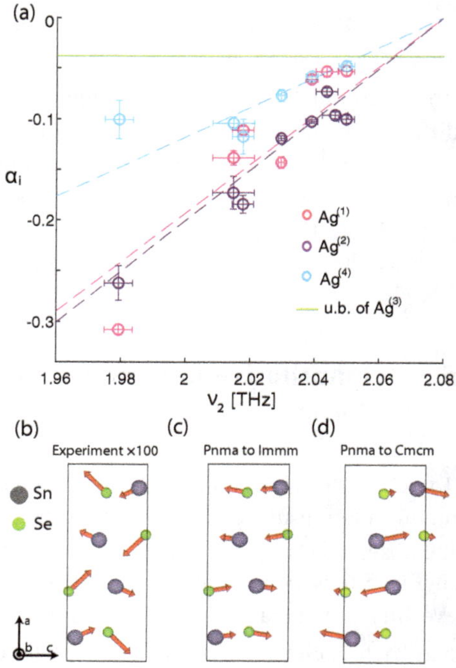

Fig. 5.7 (a) α_i as a function of $A_g^{(2)}$ frequency ν_2 and corresponding linear fits. Data points are obtained from the four Bragg peaks $(21\bar{1})$, (611), (301), and $(4\bar{1}0)$ under a range of excitation level. Lower ν_2 corresponds to higher absorbed fluence. Solid green line represents the upper bound of $|\alpha_3|$. (b) Photoexcitation-induced shift of atomic quasi-equilibrium in SnSe (red arrows) under a nominal absorbed fluence of 2.2 mJ/cm^2, magnified $\times 100$. (c), (d) Displacement connecting Pnma with Immm (c) and Cmcm (d)

are fitting parameters and the value of ν_2^0 is fixed as the equilibrium $A_g^{(2)}$ frequency 2.08 THz, which is measured with Raman spectroscopy at room temperature (see "Appendix"). The fit described above weighs in the error bars of both ν_2 and α_i.[5] We neglect $A_g^{(3)}$ for the reconstruction of the atomic motion because its amplitude upper bound (green line) is significantly smaller than other modes.[6] Summing the product of α_i and mode eigendisplacement (Fig. 5.5(c)) for all the phonon modes, one obtains the overall displacements of the quasi-equilibrium atomic positions,

[5] The error bars for ν_2 take into account only the statistical error, as in Fig. 5.5(d). The error bars for α_i take into account both the statistical error due to FEL intensity fluctuations and the systematic error. The statistical error for ν_i ($B_{hkl}^{(i)}$), as described for Fig. 5.5(d), is estimated with the standard error of the ν_i ($B_{hkl}^{(i)}$) ensemble obtained by fitting Equation (5.3) to multiple random sub-samples of the data. The uncertainty of $B_{hkl}^{(i)}$ can then map to the uncertainty of α_i. The systematic error is mainly attributed to a deviation of the grazing angle β for different azimuthal rotations. This causes slight variation in the probed volume, leading to changes in the observed signal magnitude. The β dependence of X-ray penetration depth is calculated based on [21], and the effective β is estimated to have a $\pm 0.05°$ uncertainty due to sample surface flatness. To weigh in the error bars on both x- and y-axis, one can reference, for example, [22].

[6] The upper bound of the $A_g^{(3)}$ mode amplitude α_3 is estimated from the noise level in the $(21\bar{1})$ Bragg peak data set shown in Fig. 5.5(a), taking into account the peak sensitivity (which is small but nonzero) to α_3. The noise level is estimated from the larger one of the R.M.S of the $t > 0$ fit residual (light blue trace in Fig. 5.5(a)) and the R.M.S of the $t < 0$ pre-pump noise.

described by Δx_s and Δz_s. We plot in Fig. 5.7(b) this reconstructed overall atomic displacement ($\times 100$. α_i used in the reconstruction is predicted by the linear fit in Fig. 5.7(a) at the minimum observed ν_2 value, i.e., the maximum absorbed fluence in the experiment, which is nominally 2.2 mJ/cm^2). To be specific, Fig. 5.7(b) shows $\Delta x_{\mathrm{Se}} = (1.2\pm0.1)\times10^{-3}$, $\Delta z_{\mathrm{Se}} = (-2.9\pm0.3)\times10^{-3}$, $\Delta x_{\mathrm{Sn}} = (2.6\pm1.6)\times10^{-4}$, and $\Delta z_{\mathrm{Sn}} = (2.0 \pm 0.3) \times 10^{-3}$.[7] The signs of Δx_s and Δz_s are robust within the experimental uncertainties.

5.4 Identification of a Novel Lattice Instability

To identify a new lattice structure that is compatible with a large amplitude extrapolation of the displacements presented above, it is instructive to search for higher symmetry space groups (supergroups of Pnma), since a displacement along a linear combination of A$_g$ modes cannot lower the lattice symmetry. The early time dynamics can be considered effectively at constant volume; therefore, we search for orthorhombic structures that are supergroups (higher symmetry) of Pnma, whose atomic sites can be described with the Wyckoff 4c site of Pnma for certain values of x_s and z_s after a proper coordinate transformation. We also require that the direct linear displacement toward the candidate structure must match $\Delta x_{\mathrm{Se}} > 0$, $\Delta z_{\mathrm{Se}} < 0$, $\Delta x_{\mathrm{Sn}} > 0$, and $\Delta z_{\mathrm{Sn}} > 0$ as observed in the experiment. Furthermore, we require that bond lengths d_1, d_2, d_3 do not change by more than 50% of the original length. With these criteria, we search through the 230 space groups as well as all the possible Wyckoff sites in each space group [9, 10] and find two supergroups that satisfy these constraints: Pmmn (Se:2d, Sn:2b) and Immm (Se:2d, Sn:2b). Pmmn is an intermediate structure between Pnma and Immm (see Fig. 5.2 for a group-subgroup relationship between the relevant structures). A symmetry enhancement from Pnma to Pmmn moves z_s to the high symmetry values $z_{\mathrm{Sn}} = \frac{1}{4}$ and $z_{\mathrm{Se}} = \frac{1}{4}$. A further symmetry enhancement from Pmmn toward Immm requires in addition that $x_{\mathrm{Sn}} = \frac{1}{8}$ and $x_{\mathrm{Se}} = \frac{7}{8}$ (see Table 5.1). We identify Immm as the relevant symmetry, because Pmmn does not describe the motion along **a**-axis. The conclusion is independent of the exact numerical values of Δx_s and Δz_s. The atomic displacements connecting Pnma to Immm (Cmcm) structures are plotted in Fig. 5.7(c), (d) to scale. Clearly, the signs of Δx_s, Δz_s rule out a distortion toward Cmcm as has been shown in Fig. 5.6(b).

At the same time, from Fig. 5.7(a), the magnitudes of α_1 and α_2 are significantly larger than α_3 and α_4. This is consistent with the fact that atomic displacements associated with the Pnma-Immm lattice instability (Fig. 5.7(c)) can be decomposed into restorations of high symmetry positions along the **c**- and **a**-axes, dominated by motion along the $\mathrm{A}_g^{(1)}$ and $\mathrm{A}_g^{(2)}$ coordinates, respectively. The general structural

[7] These displacements in absolute units are $a\Delta x_{\mathrm{Se}} = 1.34 \pm 0.15$ pm, $c\Delta z_{\mathrm{Se}} = -1.25 \pm 0.11$ pm, $a\Delta x_{\mathrm{Sn}} = 0.29 \pm 0.17$ pm, and $c\Delta z_{\mathrm{Sn}} = 0.85 \pm 0.13$ pm.

transformation from space group Immm to Pnma involves two irreducible representations, LD_4 (at wave vector $(0,0,\frac{1}{2})$ in the Immm Brillouin zone; see Fig. 5.4), which distorts Pnma into the Pmmn structure, and X_2^- (at wave vector (1,1,1) in the Immm Brillouin zone), which further distorts the Pmmn structure into Immm [23, 24]. In SnSe with its specific atomic sites, LD_4 describes the shear between bilayers toward a rock salt-like stacking so that Sn and Se atoms are aligned alternately along **a**-axis with high symmetry positions $z_{Sn} = \frac{1}{4}, z_{Se} = \frac{1}{4}$, while X_2^- involves the motion along **a**-axis so that Sn and Se atoms from the same atomic layer are aligned along the **c**-axis, which reduces the buckling of the bilayers in the **a**-axis and brings the atoms closer to the high symmetry positions of the Immm structure $x_{Sn} = \frac{1}{8}, x_{Se} = \frac{7}{8}$. See Fig. 5.3. Importantly, the $A_g^{(1)}$ mode of Pnma has a strong component of LD_4, and $A_g^{(2)}$ is close to X_2^-.

The total photoexcited atomic displacements reduce the difference between the d_1 and d_4 bond lengths, leading to a significant lengthening of d_1 and concomitant shortening of d_4, consistent with the Immm structure but inconsistent with the Cmcm structure (Fig. 5.1). Clearly, the softening of $A_g^{(1)}$ alone cannot distinguish between a Pnma-Cmcm and a Pnma-Immm phase transition in SnSe, highlighting the importance of ultrafast atomic-scale probes for resolving photoexcited atomic motion in materials with structural instabilities.

In Table 5.2, we provide a comparison between the experimental bond length and bond angle changes observed in the photoexcited structure and the values associated with the Pnma-Cmcm and Pnma-Immm structural distortions. The quantities $\Delta\theta$ and Δd in the table are calculated from the displacements shown in Fig. 5.7(b), without the magnification factor of $\times 100$. The bond angles are as defined in Fig. 5.8.

Table 5.2 Bond length (Å) and bond angle (°) change between structures with fixed lattice constants ($a = 11.31$ Å $b = 4.12$ Å $c = 4.30$ Å of room temperature SnSe). Pnma equilibrium structure ($\text{Pnma}_{e.q.}$) takes the fractional coordinates as calculated by DFT. Cmcm has experimentally measured fractional coordinates at 855 K. Immm fractional coordinates are as listed in Table 5.1. Bond length and angle changes for photoexcited SnSe are calculated based on Fig. 5.7(b), with error bars propagated from the linear regression error of b_i in the fit of $\alpha_i = b_i(\nu_2 - \nu_2^0)$ to data in Fig. 5.7(a). See Fig. 5.8 for definitions of $\theta_{1\text{-}4}$. The experimentally measured values with the same sign of change as Δd_{Immm} ($\Delta\theta_{\text{Immm}}$) but opposite sign of change as Δd_{Cmcm} ($\Delta\theta_{\text{Cmcm}}$) are highlighted in bold. Fractional coordinate of Cmcm at 855 K is taken from Ref. [2]

	$\text{Pnma}_{e.q.}$	$\Delta d_{p.e.} \times 100$	Immm	Δd_{Immm}	Cmcm	Δd_{Cmcm}
d_1	2.74	**1.00**(0.04)	2.83	**0.09**	2.63	**−0.11**
d_2	2.79	0.10(0.01)	2.98	0.18	2.99	0.19
d_3	3.20	−0.44(0.02)	2.98	−0.22	2.99	−0.22
d_4	3.36	**−1.94**(0.05)	2.83	**−0.53**	3.71	**0.35**
	$\text{Pnma}_{e.q.}$	$\Delta\theta_{p.e.}$	Immm	$\Delta\theta_{\text{Immm}}$	Cmcm	$\Delta\theta_{\text{Cmcm}}$
θ_1	144.62	**0.33**(0.01)	180	**35.38**	144.62	**−0.00**
θ_2	156.54	**0.12**(0.01)	180	**23.46**	144.52	**−11.92**
ϑ_3	80.37	0.37(0.02)	90	9.63	85.83	5.45
ϑ_4	99.82	−0.40(0.02)	90	−9.82	94.18	−5.64

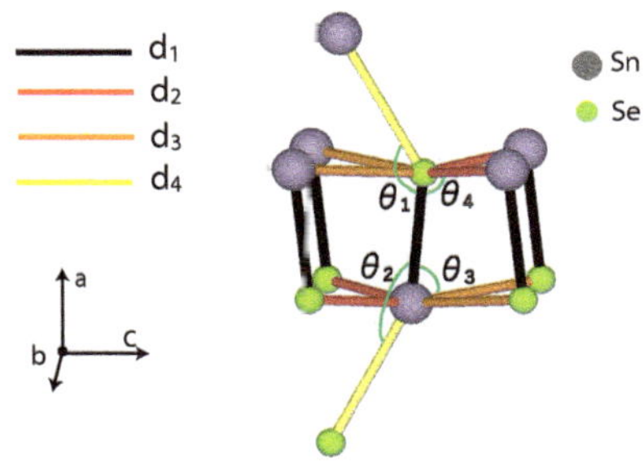

Fig. 5.8 Pnma SnSe bonds and bond angles. Atoms in the shown local structure of Pnma SnSe relative to the unit cell can be referenced in Fig. 5.1

Note that the changes toward the Immm and Cmcm phases are based on the fractional coordinates of the corresponding structures, but scaled by the Pnma lattice constant. This scaling allows for a meaningful comparison with the photoexcited state, which has constrained lattice constants. The trend to remove the corrugation within a bilayer is mainly reflected in increased θ_3 and decreased θ_4, as well as the stretched d_2 and compressed d_3 (see Table 5.2). In both Immm and Cmcm, the intralayer corrugation is reduced, explaining the same sign of Δd_2, Δd_3, $\Delta\theta_3$, and $\Delta\theta_4$ in the structural distortions toward Immm and Cmcm. Importantly, however, $\Delta d_1 > 0$, $\Delta d_4 < 0$, $\Delta\theta_1 > 0$, and $\Delta\theta_2 < 0$ are consistent with a distortion toward Immm but not with Cmcm, reflecting the fact that Immm restores the local quasi-octahedral symmetry, while Cmcm does not.

5.5 Selectivity of the Photoexcitation

To gain insight into the observed lattice instability, I worked with theory collaborators Shan Yang and Olivier Delaire from Duke University and performed constrained DFT (cDFT) as well as hole-doped DFT calculations. In cDFT, we model the effect of photoexcitation by constraining the electron occupations using two different chemical potentials for electrons and holes [25] while keeping the lattice constants fixed. cDFT calculations were performed with constrained densities $N_{eh} = 0.05$ and $N_{eh} = 0.20$ electron-hole pairs per formula unit (pairs/f.u.). For reference, from the experimental parameters, we estimate $N_{eh} = 0.15$ pairs/f.u. at 2.2 mJ/cm^2. These were estimated from the absorbed energy density per unit volume considering the reflectivity,[8] the optical penetration depth of 60 nm, and the illuminated pump area of 2 mm^2. In Fig. 5.9(a), we show the calculated x_{Sn} and z_{Sn} for several constrained charge densities N_{eh} (solid symbols in (a)). The cDFT calculations predict that Sn atoms displace toward the Immm structure instead of toward the Cmcm structure, i.e., z_{Sn} increases rather than decreases with increasing N_{eh}. The direction is consistent with the experimental results (Figs. 5.6(b) and 5.7(b)). Quantitatively, the calculated displacement $\Delta z_{\mathrm{Sn}} = 0.02$

[8] The refractive index was obtained from [26].

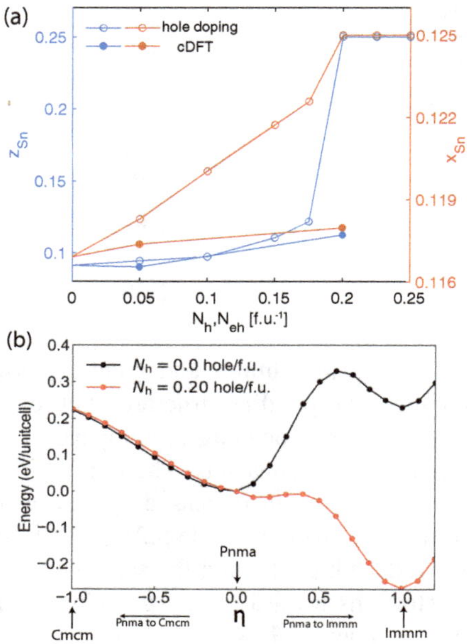

Fig. 5.9 (a) Computed x_{Sn}, z_{Sn} as a function of hole concentration, N_h (hole-doped DFT, hole/f.u.), and electron-hole concentration, N_{eh} (constrained DFT, pairs/f.u.), at constant volume. Both calculations show a tendency to distort toward the Immm structure with increasing N. For hole doping, the Immm structure ($z_{\text{Sn}} = 0.25$, $x_{\text{Sn}} = 0.125$) is obtained at and above $N_h = 0.2$ hole/f.u. (b) Potential energy as a function of atomic configurations interpolated between Cmcm-Pnma ($-1 \leq \eta \leq 0$) and Pnma-Immm ($0 \leq \eta \leq 1$), at different hole doping levels $N_h = 0, 0.2$ hole/f.u. The Immm structure is stabilized at $N_h = 0.2$ hole/f.u, evidenced by the energy minimum at $\eta = 1$

is an order of magnitude larger than the measured displacement $\Delta z_{\text{Sn}} = 2.0 \times 10^{-3}$ for 0.15 pairs/f.u.

In addition to the cDFT calculations, a simplified version of DFT calculations was performed by introducing hole doping and fixing the lattice constants. This simplified approach captures the essential features of the distortion and exhibits better numerical convergence compared to cDFT. In these calculations, the structure of SnSe was found to distort from the Pnma phase toward the higher symmetry Immm phase with increasing hole density.

The hole doping DFT calculations predict an abrupt phase transition to the Immm structure at a hole concentration of approximately 0.2 hole/f.u., as shown in Fig. 5.9(a). This transition is accompanied by a suppression of the 0.33 eV/unit-cell energy barrier near $\eta = 0.5$ for the case of no hole doping ($N_h = 0$ hole/f.u.). The investigated hole doping levels of $N_h = 0$ and 0.2 hole/f.u. do not significantly affect the energy landscape between the Pnma and Cmcm structures. Specifically, the energy of the Cmcm structure remains higher than that of Pnma even under hole doping. In Fig. 5.9(b), the calculated total energy for a series of configurations

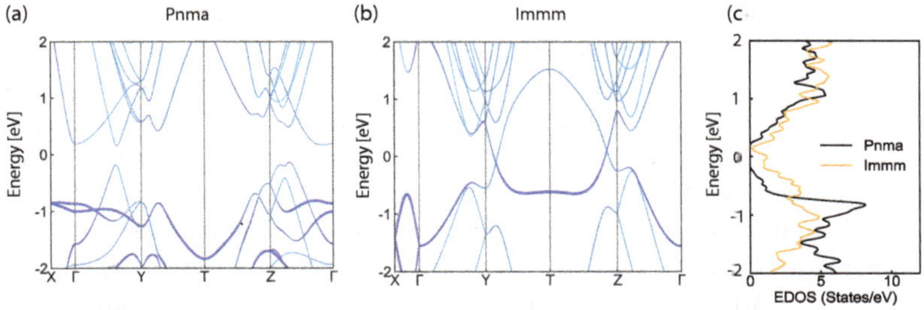

Fig. 5.10 Calculated electronic band structure of (a) equilibrium Pnma phase (N_h = 0.0 hole/f.u.) and (b) hole-doped Immm structure (N_h = 0.20 hole/f.u.). The blue line thickness represents the band projection onto the Se $4p_x$ orbital. (c) Electron density of states of Pnma (N_h = 0.0 hole/f.u.) and Immm (N_h = 0.20 hole/f.u.) structures

between Pnma-Immm and Pnma-Cmcm is shown for two different hole concentrations, $N_h = 0$ and 0.2 hole/f.u. The parameter η parameterizes the structural configuration, representing linear interpolations between Pnma ($\eta = 0$), Immm ($\eta = +1$), and Cmcm ($\eta = -1$). The abrupt structural phase transition upon hole doping at $N_h = 0.2$ hole/f.u. is associated with a suppression of the energy barrier near $\eta = 0.5$ for $N_h = 0$ hole/f.u. However, the investigated hole doping levels do not significantly affect the energy landscape between the Pnma and Cmcm structures. Notably, the energy of the Cmcm structure remains higher than that of Pnma even under hole doping. The presence of an energy minimum at $\eta = 1$ under 0.2 hole/f.u. suggests that the Immm structure may be realized at sufficiently high excitation densities if sample damage can be mitigated. These theoretical findings provide further insight into the effect of photoexcitation and the associated energy landscape in SnSe.

According to the hole-doped DFT calculations, the photoinduced Immm structural instability in SnSe can be attributed to the excitation of electrons out of the Se $4p_x$ orbital-derived bands through the absorption of NIR photons. Figure 5.10 displays the calculated electronic band structure of SnSe in the Pnma phase (Fig. 5.13(a)) and the transient photoexcited Immm phase (Fig. 5.13(b)) with a hole doping level of $N_h = 0.2$ hole/f.u. The Brillouin zone labels follow the Pnma unit cell convention (refer to Fig. 5.4). In both (a) and (b), the Se $4p_x$ orbital character is represented by the thick portions of the bands. In the Pnma phase, the Se $4p_{y,z}$ orbitals hybridize with Sn 5s orbitals and form a symmetric resonantly bonded network in the $\mathbf{b} - \mathbf{c}$ plane. This hybridized orbital is the main character of the edge of the valence bands and contributes to the Pnma-Cmcm phase transition through a Peierls-like mechanism [5, 19]. The Se $4p_x$ orbital, however, is hybridized with Sn 5s to form the non-dispersive band along $\Gamma - X$ about 0.7 eV below the top of the valence band in Fig. 5.10(a). From the electron density of states (EDOS) in Pnma phase (Fig. 5.10(c)), the holes will populate down to -0.77 eV under 0.2 hole/f.u assuming holes are filled from the top of the valence band, and −0.77 eV is close to the peak of the EDOS, which is mostly formed by the non-dispersive bands due to non-bonding lone pairs in Pnma (compare Fig. 5.10(a) and (c)). Removal of electrons from these non-dispersive bands causes suppression of the lone pair stereo-chemical activity, which is generally considered [27] to raise the structural symmetry, in our case, the symmetry of the local quasi-octahedral coordination. In fact, in the Immm structure, the electron bands feature a clear dispersion of the band consisting mainly of Se $4p_x$ and Sn $5s$ orbital components (compare Fig. 5.10(a) and (b) between $\Gamma - X$), which reflects the disappearance of the non-bonding localized lone pair orbitals.

Figure 5.10(a) and (b) show that along $T - Y$ and $T - Z$ in Immm, the Se p_x bands and Sn p_x bands are inverted and the bandgaps that exist in Pnma close.[9] The band crossings along $T - Y$ and $T - Z$ do not occur at the same energy due to orthorhombic structural distortion, leading to a finite EDOS everywhere in Fig. 5.10. The Pnma-Immm structural instability and its connection to the disappearance of the lone pair are reminiscent of the structural phases formed by other group V or IV–VI rock salt-derived materials [28, 29]. The photoinduced structural instability has a Peierls mechanism nature but originates from different orbitals than the Pnma-Cmcm instability. Though the band inversion and gap closing along $T - Y$ and $T - Z$ in SnSe Immm bear a resemblance to the electron band dispersion in the rock salt TCI of IV–VI compounds [30–32], Immm lacks the proper lattice symmetry to become a TCI. See Chap. 3.

In conclusion, our study reveals that ultrafast near-infrared (NIR) photoexcitation of SnSe leads to a structural instability toward the Immm phase, an orthorhombically distorted rock salt structure, rather than the thermodynamically stable Cmcm phase. This photoinduced structural instability is distinct from the thermal transition and involves different electron orbitals. Our density functional theory (DFT) calculations indicate that the excitation of electrons from non-bonding Se $4p_x$ orbitals to Sn $5s$ orbitals by the 1.55 eV laser pulse is responsible for the Immm instability. We note that alloying with Pb may reduce the Immm phase transition threshold [33–35], while lowering the temperature will increase the damage threshold (the experiments reported here were limited in the maximum carrier density we could produce without damaging the sample), such that a photoinduced Immm phase may be realizable. Our findings may also have implications in other rock salt distorted IV–VI semiconductors, several of which have topological states protected by lattice symmetry in the cubic or tetragonal phases [30, 31, 36, 37]. More generally, our work suggests that pump wavelength could provide additional control of structural distortions through orbitally selective above-gap excitation. This could be exploited to direct a particular structural distortion to desirable outcomes with particular functionality beyond those accessible in thermal equilibrium.

9 $T - Y$ and $T - Z$ correspond to $W - S$ and $W - R$ in Immm convention (see Fig. 5.4), which would correspond to $W - L$ in the Fm$\bar{3}m$ Brillouin zone for the rock salt structure.

Appendix

Raman Spectroscopy and Pump-Probe Reflectivity Measurements

The room temperature Raman measurements were carried out using a Horiba LabRAM HR Evolution spectrometer equipped with a continuous-wave laser operating at a photon energy of 1.96 eV (wavelength 633 nm). The Raman spectrum was acquired under the $x(yy)\bar{x}$ and $x(zz)\bar{x}$ geometries, where the incident and reflected beams have the same polarization. For the room temperature pump-probe reflectivity measurements, a Coherent RegA Ti:sapphire laser system was utilized. The laser system operated at a repetition rate of 250 kHz with a photon energy of 1.55 eV (800 nm) and a pulse duration of 46 fs for both the pump and probe pulses (Fig. 5.11). The pump-probe measurement was performed under normal incidence, with the incoming beam and reflected beam propagating along the crystal **a**-axis. Both the pump and probe pulses were polarized along either the **b**-axis or the **c**-axis. The same single crystal sample used in the time-resolved X-ray diffraction experiment was also employed for both the spontaneous Raman and pump-probe measurements.

Figure 5.12(a) and (b) show the time-resolved optical reflectivity of SnSe with polarizations along **b** and **c**. (c) and (d) show the Fourier transforms of (a) and (b), as well as the Raman spectrum under $x(yy)\bar{x}$ and $x(zz)\bar{x}$. Our Raman spectrum is consistent with previous measurements [11].

The frequency of the $A_g^{(2)}$ mode measured in Raman spectroscopy is 2.08 THz. If we allow the ν_2^0 to be a fitting parameter in a global fit $\alpha_i = b_i(\nu_2 - \nu_2^0)$, $i \in \{1, 2, 4\}$ to data in Fig. 5.7(a), $\nu_2^0 \approx 2.07$ THz. The $A_g^{(3)}$ mode around 3.8 THz is clear in these optical measurements, while it was not detected above the noise in time-resolved X-

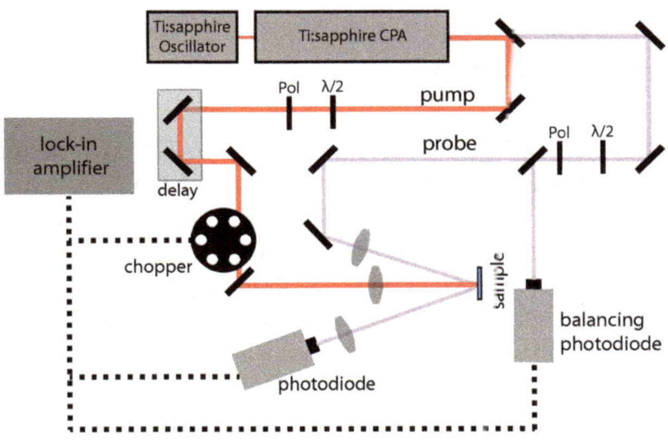

Fig. 5.11 Time-resolved optical reflectivity setup schematic

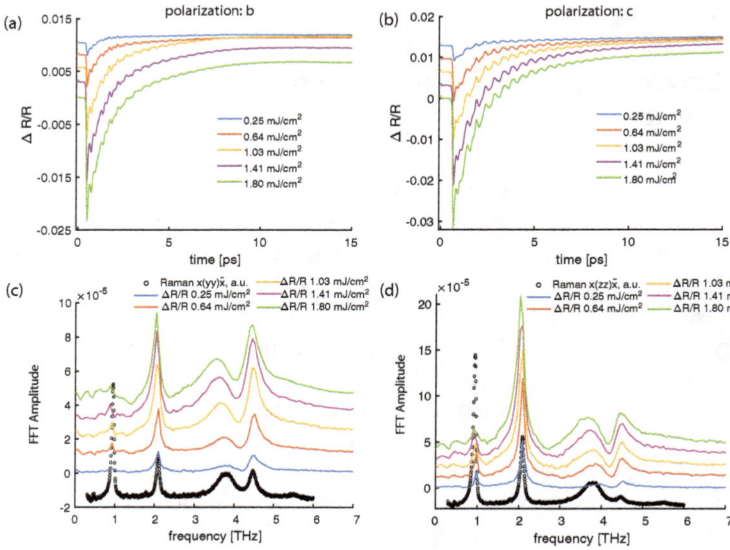

Fig. 5.12 Raman spectrum and optical reflectivity. (a), (b) Optical reflectivity data, with both pump and probe beam polarized along **b** and **c**, respectively. The fluence range selected for reflectivity measurements matches that of the time-resolved X-ray diffraction experiment. (c), (d) Raman spectrum under $x(yy)\bar{x}$ and $x(zz)\bar{x}$ geometries (black circles), shown together with Fourier transform of pump-probe reflectivity data in (a) and (b) (colored lines). All four A_g modes of SnSe are identified in both Raman measurement and pump-probe optical reflectivity

ray diffraction data. The drastic softening of $A_g^{(3)}$, a mode that is mainly polarized in the $\mathbf{b}-\mathbf{c}$ plane (Fig. 5.5), can be related to the strengthening of the resonant bonds that also result in the large softening of $A_g^{(1)}$ as observed in the diffraction results.

Density Functional Theory

DFT (and hole-doped DFT) was performed using VASP, with the projector augmented wave (PAW) and local density approximation (LDA) [38–40], which proves to yield accurate phonon dispersions [5, 20, 41] and provides better agreement with INS and Raman measurements than the Perdew-Burke-Ernzerhof (PBE) generalized gradient approximation (GGA) [20, 41]. The Pnma equilibrium structure is relaxed with a kinetic energy cutoff of 500 eV and an electronic Monkhorst-Pack grid with $6 \times 12 \times 12$ k-points, giving lattice constants ($a = 11.31\,\text{Å}$ $b = 4.12\,\text{Å}$ $c = 4.30\,\text{Å}$) in good agreement with X-ray diffraction experimental report at 296 K: [3] ($a = 11.50\,\text{Å}$ $b = 4.16\,\text{Å}$ $c = 4.45\,\text{Å}$) and [2] ($a = 11.44\,\text{Å}$ $b = 4.13\,\text{Å}$ $c = 4.45\,\text{Å}$). We use the DFT calculated fractional coordinates ($x_{\text{Se}} = 0.14$, $z_{\text{Se}} = 0.48$, $x_{\text{Sn}} = 0.88$, $z_{\text{Sn}} = 0.09$) for the equilibrium Pnma structure, which are also

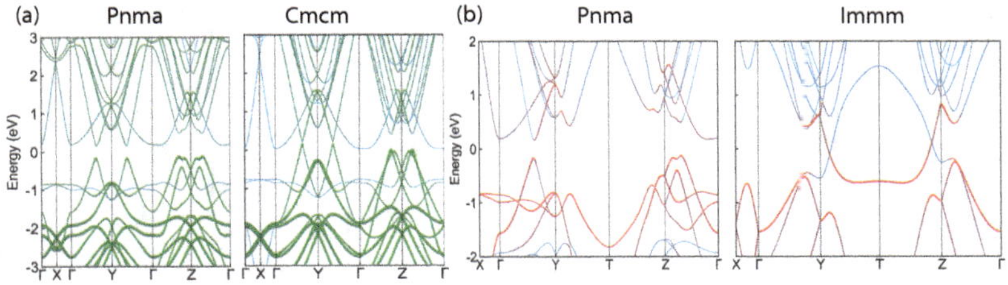

Fig. 5.13 (a) Equilibrium Pnma phase and Cmcm phase electron band structure. Thickness of green lines represents the projection to Se $p_{y,z}$ orbitals. (b) Equilibrium Pnma phase ($N_h = 0.0$ hole/f.u.) and photoexcited Immm structure ($N_h = 0.20$ hole/f.u.) electron band structure. Thickness of red lines represents the projection to Sn 5s orbitals

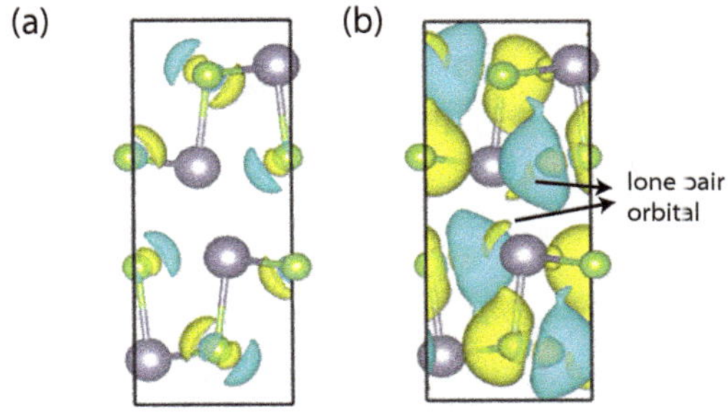

Fig. 5.14 Photoexcited state differential charge isosurface (0.022 electron/Å^3) plot, for $N_h = 0.175$ hole/f.u. (a) and for $N_h = 0.20$ hole/f.u. (b) Atoms in the unit cells are of the Pnma equilibrium structure. The yellow isosurface represents a differential (photoexcited subtracted by equilibrium) positive charge (negative EDOS), while blue isosurface represents a negative charge difference (positive EDOS)

in good agreement with [2] ($x_{\mathrm{Se}} = 0.12$, $z_{\mathrm{Se}} = 0.48$, $x_{\mathrm{Sn}} = 0.86$, $z_{\mathrm{Sn}} = 0.11$). In constrained DFT, the structure optimization was performed on the $2 \times 4 \times 4$ supercell of Pnma conventional unit cell, using a $3 \times 3 \times 3$ k-point mesh and 500 eV energy cutoff. In both calculation methods, the lattice constants were fixed to those of the equilibrium Pnma structure.

Figure 5.13(a) shows Se 4$p_{y,z}$ orbital projected bands in both Pnma and Cmcm phases. The bandgap is closed along $\Gamma - Y$ by bands mainly composed of Se 4$p_{y,z}$ (hybridized with Sn 5s) in a phase transition from Pnma to Cmcm, which reveals the Peierls nature of the lattice instability [5, 19]. The non-bonding lone pair orbital is a hybridized orbital of Sn 5s and Se 4p_x; the electron bands' projections onto the latter are shown in the main text. For reference, we also plot the Sn 5s orbital projected Pnma and Immm band structure in Fig. 5.13(b).

Below the excitation threshold $N_h = 0.20$ hole/f.u., the electron density isosurface plot in Fig. 5.14(a) ($N_h = 0.175$ hole/f.u.) features hole doping into the $\mathbf{b} - \mathbf{c}$ plane p orbitals that are closer to the Fermi level than the lone pair orbitals. The electron density change of lone pair orbital (around the Sn atom in Fig. 5.14v) is manifest at

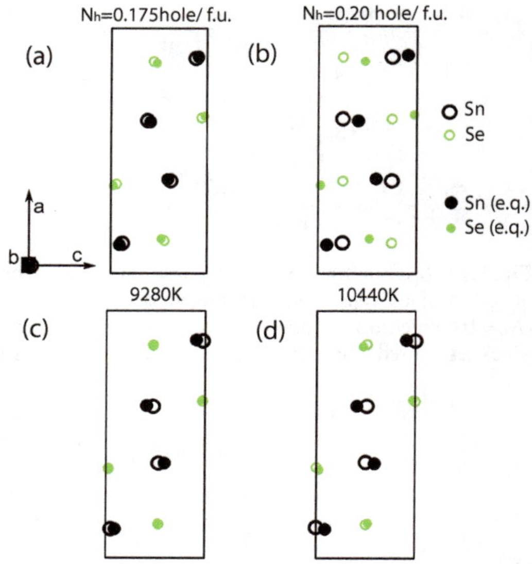

Fig. 5.15 Comparison between hole-doped DFT and DFT with modified electronic temperatures. The Sn (Se) atoms are empty gray (green) circles for structures with modified electronic structures and filled circles for the equilibrium. (a, b) Crystal structures with the increasing number of holes per SnSe formula unit N_h. Atomic positions are displaced toward Immm as the excitation density increases. (c, d) DFT calculations by raising the electronic temperatures as indicated with the title for each plot, showing atomic displacements toward Cmcm as the temperatures increase

a higher excitation density $N_h = 0.20$ hole/f.u., which has a smaller isosurface than the $\mathbf{b} - \mathbf{c}$ plane p orbitals but a larger effect in determining the structural distortion.

We note that DFT calculations assuming increased electronic temperature and the same chemical potential for both electrons and holes will leave the electrons occupying the lone pair orbitals mostly intact and do not even qualitatively reproduce the experimentally observed atomic motion (see Fig. 5.15). This suggests that the theoretical formalism for non-equilibrium photoexcited material needs to be dealt with care to predict material behavior under above-gap excitation. This effort will likely need to be combined with microscopic experimental probes.

Linear Prediction and Phonon Mode Decomposition

Linear prediction decomposes the data into a sum of exponentially decaying harmonic oscillators and decaying exponentials [12]. It has been applied to the analysis of time-domain data, such as NMR spectra [12, 42]. Unlike nonlinear least squares, this method does not require the initial guess of fitting parameters. It is a convex optimization problem and, thus, will not be stuck in a local minimum as in a least-square fit. Furthermore, linear prediction can give a statistically sound estimate of the number of oscillators contained in the signal [43, 44], which is usually an external input in a least-square fit. Similar to Fig. 5.6(a) for $(21\bar{1})$ Bragg peak, Fig. 5.16(a) and (b) show the decomposition prescribed by Equation (5.3) for the (611) and (301) Bragg peaks. The black trace is the experimental data, and the colored lines are the DECP components. In (301), the $A_g^{(4)}$ component is not visible.

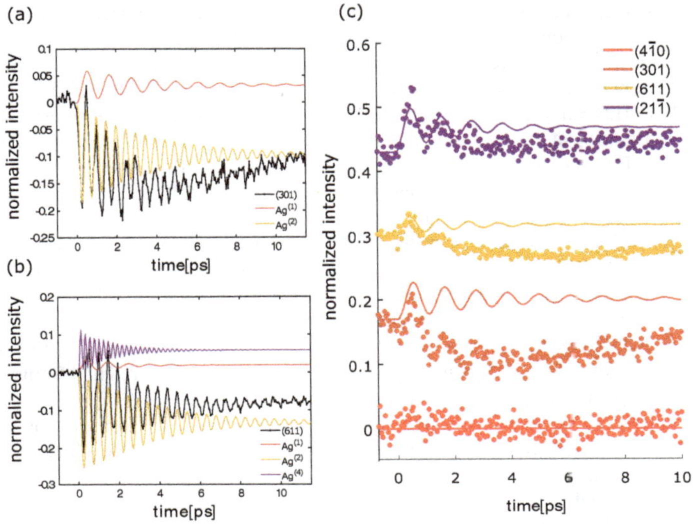

Fig. 5.16 (a) Linear prediction showing decomposition of modes for peak (611) and similarly (b) for peak (301). The black traces are the raw data, and the colored traces are the components of A_g modes. (c) The color dots represent the residual of raw data after subtracting off the $A_g^{(2)}$ and $A_g^{(4)}$ components; the solid color lines represent the $A_g^{(1)}$ component

Figure 5.16(c) shows an analysis of the isolated $A_g^{(1)}$ component for all four peaks $(21\bar{1})$, (611), (301), and $(4\bar{1}0)$. Here, we show the residual (colored dots) of the experimental intensity subtracted by the components of $A_g^{(2)}$ and $A_g^{(4)}$, as well as the linear predicted $A_g^{(1)}$ component (colored lines). The residual is, of course, very noisy since modes with high signal levels are subtracted. Nonetheless, the initial phase of the $A_g^{(1)}$ is manifest in both the residual and the linear predicted DECP component. We note that peak $(4\bar{1}0)$ is not sensitive to mode $A_g^{(1)}$ since the motion is primarily along the z direction.

References

1. Y. Huang, S. Yang, S. Teitelbaum, G. De la Peña, T. Sato, M. Chollet, D. Zhu, J.L. Niedziela, D. Bansal, A.F. May, A.M. Lindenberg, O. Delaire, D.A. Reis, M. Trigo, Observation of a novel lattice instability in ultrafast photoexcited SnSe. Phys. Rev. X **12**, 011029 (2022)
2. M. Sist, J. Zhang, B.B. Iversen, Crystal structure and phase transition of thermoelectric SnSe. Acta Crystallographica Sect. B: Struct. Sci. Crystal Eng. Mater. **72**(3), 310–316 (2016)
3. H. Wiedemeier, F.J. Csillag, The thermal expansion and high temperature transformation of SnS and SnSe. Zeitschrift für Kristallographie – Crystalline Materials **149**(1–4), 17–30 (1979)
4. K. Momma, F. Izumi, VESTA: a three-dimensional visualization system for electronic and structural analysis. J. Appl. Crystallogr. **41**(3), 653–658 (2008)

5. C.W. Li, J. Hong, A.F. May, D. Bansal, S. Chi, T. Hong, G. Ehlers, O. Delaire, Orbitally driven giant phonon anharmonicity in SnSe. Nat. Phys. **11**(12), 1063 (2015)
6. T. Chattopadhyay, J. Pannetier, H.G. Von Schnering, Neutron diffraction study of the structural phase transition in SnS and SnSe. J. Phys. Chem. Solids **47**(9), 879–885 (1986)
7. I. Loa, R.J. Husband, R.A. Downie, S.R. Popuri, J.-W.G. Bos, Structural changes in thermoelectric SnSe at high pressures. J. Phys.: Condens. Matter **27**(7), 072202 (2015)
8. S. Ivantchev, E. Kroumova, G. Madariaga, J.M. Perez-Mato, M.I. Aroyo, Subgroupgraph: a computer program for analysis of group–subgroup relations between space groups. J. Appl. Crystallogr. **33**(4), 1190–1191 (2000)
9. M.I. Aroyo, J.M. Perez-Mato, C. Capillas, E. Kroumova, S. Ivantchev, G. Madariaga, A. Kirov, H. Wondratschek, Bilbao crystallographic server: I. Databases and crystallographic computing programs. Zeitschrift für Kristallographie-Crystalline Materials **221**(1), 15–27 (2006)
10. M.I. Aroyo, J.M. Perez-Mato, D. Orobengoa, E. Tasci, G. De La Flor, A. Kirov, Crystallography online: Bilbao crystallographic server. Bulgarian Chem. Commun. **43**(2), 183–197 (2011)
11. H.R. Chandrasekhar, R.G. Humphreys, U. Zwick, M. Cardona, Infrared and Raman spectra of the IV-VI compounds SnS and SnSe. Phys. Rev. B **15**, 2177–2183 (1977)
12. H. Barkhuijsen, R. de Beer, W.M.M.J. Bovee, D. van Ormondt, Retrieval of frequencies, amplitudes, damping factors, and phases from time-domain signals using a linear least-squares procedure. J. Mag. Reson. (1969) **61**(3), 465–481 (1985)
13. H.J. Zeiger, J. Vidal, T.K. Cheng, E.P. Ippen, G. Dresselhaus, M.S. Dresselhaus, Theory for displacive excitation of coherent phonons. Phys. Rev. B **45**, 768–778 (1992)
14. S.W. Teitelbaum, T.C. Henighan, H. Liu, M.P. Jiang, D. Zhu, M. Chollet, T. Sato, É.D. Murray, S. Fahy, S. O'Mahony, T.P. Bailey, C. Uher, M. Trigo, D.A. Reis, Measurements of nonequilibrium interatomic forces using time-domain x-ray scattering. Phys. Rev. B **103**, L180101 (2021)
15. R. Merlin, Generating coherent THz phonons with light pulses. Solid State Commun. **102**(2), 207–220 (1997). Highlights in Condensed Matter Physics and Materials Science
16. D.A. Reis, A.M. Lindenberg, *Ultrafast x-Ray Scattering in Solids* (Springer, Berlin, 2006)
17. M. Born, K. Huang, *Dynamical Theory of Crystal Lattices* (Clarendon Press, Oxford, 1954)
18. N.W. Ashcroft, N.D. Mermin et al., *Solid State Physics* (Holt, Rinehart and Winston, New York/London, 1976)
19. J. Hong, O. Delaire, Phase transition and anharmonicity in SnSe. Mater. Today Phys. **10**, 100093 (2019)
20. T. Lanigan-Atkins, S. Yang, J.L. Niedziela, D. Bansal, A.F. May, A.A. Puretzky, J.Y.Y. Lin, D.M. Pajerowski, T. Hong, S. Chi, G. Ehlers, O. Delaire, Extended anharmonic collapse of phonon dispersions in SnS and SnSe. Nat. Commun. **11**(1), 1–9 (2020)
21. B.L. Henke, E.M. Gullikson, J.C. Davis, X-ray interactions: photoabsorption, scattering, transmission, and reflection at E = 50–30,000 eV, Z = 1–92. Atomic Data Nucl. Data Tables **54**(2), 181–342 (1993)
22. W.H. Press, S.A. Teukolsky, Fitting straight line data with errors in both coordinates. Comput. Phys. **6**(3), 274–276 (1992)
23. L. Elcoro, B. Bradlyn, Z. Wang, M.G. Vergniory, J. Cano, C. Felser, B.A. Bernevig, D. Orobengoa, G. Flor, M.I. Aroyo, Double crystallographic groups and their representations on the bilbao crystallographic server. J. Appl. Crystallogr. **50**(5), 1457–1477 (2017)
24. M.I. Aroyo, A. Kirov, C. Capillas, J.M. Perez-Mato, H. Wondratschek, Bilbao crystallographic server. II. Representations of crystallographic point groups and space groups. Acta Crystallogr. Sect. A: Found. Crystallogr. **62**(2), 115–128 (2006)
25. P. Tangney, S. Fahy, Density-functional theory approach to ultrafast laser excitation of semiconductors: application to the A_1 phonon in tellurium. Phys. Rev. B **65**, 054302 (2002)
26. O. Madelung, U. Rössler, M. Schulz (eds.), *Tin selenide (SnSe) Optical Properties, Dielectric Constants* (Springer, Berlin/Heidelberg, 1998)
27. L.E. Orgel, The stereochemistry of B subgroup metals. Part II. The inert pair. J. Chem. Soc. 3815–3819 (1959). https://doi.org/10.1039/JR9590003815

28. W. Tremel, R. Hoffmann, Tin sulfide, tellurium iodide ((Te2)2I2), and related compounds: symmetry-controlled deformations in solid-state materials. Inorganic Chem. **26**(1), 118–127 (1987)
29. J. Galy, G.L. Vignoles, The role of p $3s^2$ lone pair (e) in structure, properties and phase transitions of black phosphorus. stereochemistry and ab initio topology analyses. Solid State Sci. **100**, 106068 (2020)
30. T.H. Hsieh, H. Lin, J. Liu, W. Duan, A. Bansil, L. Fu, Topological crystalline insulators in the SnTe material class. Nat. Commun. **3**(1), 1–7 (2012)
31. S.-Y. Xu, C. Liu, N. Alidoust, M. Neupane, D. Qian, I. Belopolski, J.D. Denlinger, Y.J. Wang, H. Lin, L.A. Wray et al., Observation of a topological crystalline insulator phase and topological phase transition in $Pb_{1-x}Sn_xTe$. Nat. Commun. **3**(1), 1–11 (2012)
32. Y. Sun, Z. Zhong, T. Shirakawa, C. Franchini, D. Li, Y. Li, S. Yunoki, X.-Q. Chen, Rocksalt SnS and SnSe: Native topological crystalline insulators. Phys. Rev. B **88**, 235122 (2013)
33. P.B. Littlewood, The crystal structure of IV-VI compounds. I. Classification and description. J. Phys. C: Solid State Phys. **13**(26), 4855–4873 (1980)
34. T. Inoue, H. Hiramatsu, H. Hosono, T. Kamiya, Nonequilibrium rock-salt-type Pb-doped SnSe with high carrier mobilities 300 $cm^2/(Vs)$. Chem. Mater. **28**(7), 2278–2286 (2016)
35. K. Behnia, Finding merit in dividing neighbors. Science **351**(6269), 124–124 (2016)
36. P. Dziawa, B.J. Kowalski, K. Dybko, R. Buczko, A. Szczerbakow, M. Szot, E. Łusakowska, T. Balasubramanian, B.M. Wojek, M.H. Berntsen et al., Topological crystalline insulator states in $pb_{1-x}sn_xse$. Nat. Mater. **11**(12), 1023–1027 (2012)
37. Y. Tanaka, Z. Ren, T. Sato, K. Nakayama, S. Souma, T. Takahashi, K. Segawa, Y. Ando, Experimental realization of a topological crystalline insulator in SnTe. Nat. Phys. **8**(11), 800–803 (2012)
38. G. Kresse, J. Furthmüller, Efficient iterative schemes for ab initio total-energy calculations using a plane-wave basis set. Phys. Rev. B **54**, 11169–11186 (1996)
39. P.E. Blöchl, Projector augmented-wave method. Phys. Rev. B **50**, 17953–17979 (1994)
40. G. Kresse, D. Joubert, From ultrasoft pseudopotentials to the projector augmented-wave method. Phys. Rev. B **59**, 1758–1775 (1999)
41. D. Bansal, J. Hong, C.W. Li, A.F. May, W. Porter, M.Y. Hu, D.L. Abernathy, O. Delaire, Phonon anharmonicity and negative thermal expansion in SnSe Phys. Rev. B **94**, 054307 (2016)
42. J.J. Led, H. Gesmar, Application of the linear prediction method to NMR spectroscopy. Chem. Rev. **91**(7), 1413–1426 (1991)
43. B.P. Epps, E.M. Krivitzky, Singular value decomposition of noisy data: noise filtering. Exp. Fluids **60**(8), 1–23 (2019)
44. B.P. Epps, E.M. Krivitzky, Singular value decomposition of noisy data: mode corruption. Exp. Fluids **60**(8), 1–30 (2019)

Chapter 6
Nonthermal Bonding Origin of the Novel Lattice Instability

Abstract We present time-resolved X-ray scattering measurements that provide decisive information about the nonequilibrium lattice dynamics in SnSe. By analyzing the excited-state phonon dispersions obtained from these measurements and fitting them to an interatomic force model, we identify that the change in interatomic interaction of the bilayer-connecting d_4 bond is responsible for destabilizing the photoexcited SnSe structure and causing soft phonons. This highlights the significance of studying nonequilibrium lattice dynamics alongside diffraction techniques, as the former provides insights into interatomic bonding beyond the average lattice structure.

In this chapter, part of the figures are reproduced from Ref. [1] with permission from American Physical Society. DFT calculations are provided by collaborators Shan Yang and Olivier Delaire at Duke University. The last section of Appendix contains unpublished data.

6.1 Probing the Excited-State Lattice Dynamics

The interatomic forces that determine materials' structure and dynamics can be modified by temperature, pressure, chemical composition, and applied fields leading to new equilibrium phases with dramatically different electronic, thermal, and mechanical properties [2–4]. Methods such as inelastic X-ray [5–7] and neutron scattering [8, 9] are valuable because they can measure how the lattice dynamics change upon approaching or undergoing a phase transition. Ultrafast photoexcitation can also be used to generate nonequilibrium states on a timescale that is short compared to various relaxation pathways. In some cases the excitation is only a small perturbation, and much can be learned about the equilibrium dynamics by studying how the material returns to equilibrium. In other cases, novel states of

Y. Huang, *Towards the Optical Control of Resonantly Bonded Materials*, Springer Theses, https://doi.org/10.1007/978-3-031-42826-5_6

matter may be produced, with properties that do not exist in equilibrium [10]. Either way, conventional lattice dynamics measurements are not able to access the short timescales involved in the transition, and thus we need new tools for measuring transient, excited-state phonon dispersion. Time-domain, femtosecond X-ray diffuse scattering from free-electron lasers has recently been demonstrated to be able to give access to the phonon dispersion following ultrafast photoexcitation, by measuring temporal coherences in the mean square atomic displacements as a function of momentum transfer, associated with a rapid change in the interatomic forces [11–13]. In the case of photoexcited bismuth, Teitelbaum et al. [14] were able to measure changes in the interatomic forces associated with the well-established partial reversal of the Peierls distortion upon above gap excitation. Here we extend this technique to the study of a novel photoinduced lattice instability in the resonantly bonded monochalogenide, SnSe. As discussed in Chap. 5, we have recently found through ultrafast X-ray diffraction that SnSe undergoes a lattice instability towards a higher symmetry structure that is distinct from its well-known high-temperature phase transition that is associated with the materials' low thermal conductivity and high thermoelectric figure of merit [15]. The new instability drives the material towards a structure that is an orthorhombic distortion of the cubic rock salt structure which is important for its topological properties. While time-resolved X-ray diffraction can yield changes in the average structure of the material, it does not directly yield microscopic information on the changes in bonding, or the electronic states, that give rise to it.

Time-resolved diffuse scattering [11] requires a high X-ray photon flux of FELs, as diffuse scattering efficiency is orders of magnitude lower than Bragg peaks. The pump pulse is centered at 1.55 eV and the probe beam is 9.5 keV X-ray. The X-ray beam impinges on the SnSe sample at grazing incidence to better match the pump and probe, given the large disparity in their absorption. The crystal is oriented by rotating its azimuth angle φ, and a two-dimensional detector allows us to map out a large portion of the Ewald sphere. See Chap. 4, Fig. 4.4, for the setup schematic.

In Fig. 6.1(a), we show the time dependence of diffuse scattering intensity for selected scattering vectors (momentum transfer) $\mathbf{Q} = (H1\bar{1})$ where $H \in [3, 4]$

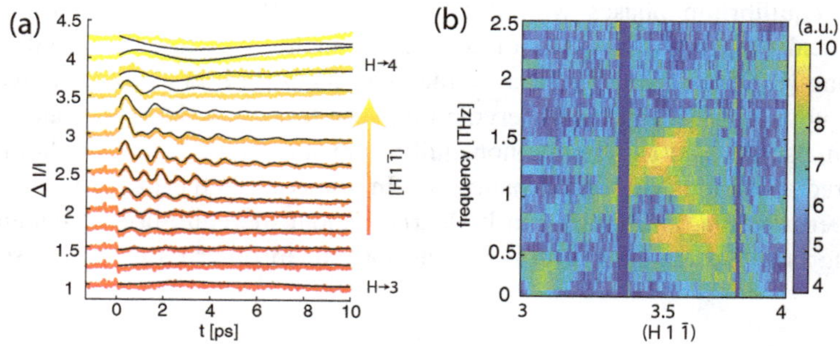

Fig. 6.1 Time-resolved diffuse scattering of SnSe. (a) Time dependence of the relative intensity for representative $\mathbf{Q}$ along $(H1\bar{1})$, $H \in [3,4]$. Black lines show reconstructions based on linear prediction. (b) The magnitude of the Fourier transform of time traces as those shown in (a)

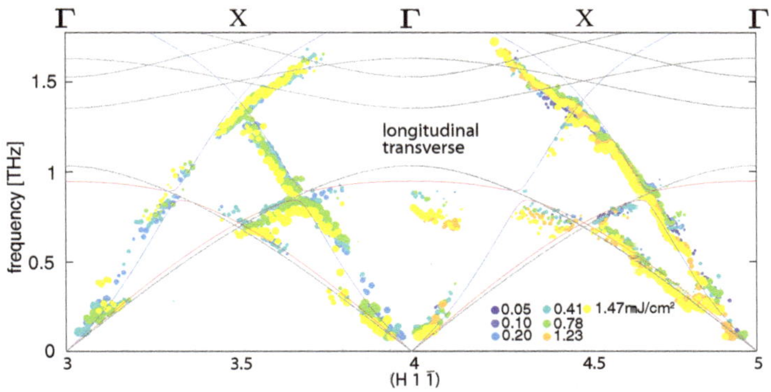

Fig. 6.2 Photoexcited phonon dispersion of SnSe. Low-frequency region of the dispersion along for phonons propagating along the a-direction ($\mathbf{Q} = (H1\bar{1})$, $H \in [3,5]$) extracted from linear prediction (see text). The solid lines are the ground-state phonon dispersion based on density functional theory calculations. Red and blue lines represent the lowest two transverse and longitudinal branches polarized along **c**-axis and **a**-axis, respectively

in reciprocal lattice units of the orthorhombic Pnma structure. The scattered X-ray intensity is modulated as a function of the pump-probe delay t. Figure 6.1(b) is a color plot showing the magnitude of the Fourier transform of time-domain data as shown in Fig. 6.1(a) and clearly shows dispersive modes. We extract the frequencies of oscillations by a linear prediction (LP) method that decomposes the data into a sum of decaying cosines where the number of oscillators is determined directly from the data [16–18]. The black lines in Fig. 6.1(a) are the sum of the LP components of the data, which correspond to phonon modes.

Figure 6.2 combines data collected over the reciprocal space range $\mathbf{Q} = (H1\bar{1})$, ($H \in [3, 5]$). The size of the dots represents the log-scaled amplitude of the oscillations from the LP, while the colors of the dots represent different optical pump fluences. The solid lines show the phonon dispersion computed from density functional theory (DFT)-based calculations, which reproduce well those from inelastic neutron scattering [19–21]. The branches shown in red are the **c**-polarized transverse acoustic (TA) branch which folds into the lowest transverse optical (TO) branch. From here on, these two branches are referred together as $T(\mathbf{c})$. Similarly, the blue line shows the **a**-polarized longitudinal acoustic (LA) branch that folds into the lowest longitudinal optical (LO) branch (referred together as $L(\mathbf{a})$). The assignment of $L(\mathbf{a})$ and $T(\mathbf{c})$ phonons branches with reduced wave vectors $\mathbf{q} = (h00)$ (along $\mathbf{\Gamma}(h = 0) - \mathbf{X}(h = 0.5)$) is based on the polarization selectivity of the phonon structure factor. We do not observe any oscillations in the diffuse scattering attributed to **b**-polarized modes, and for clarity, we do not include their calculated dispersion here.

The entire $T(\mathbf{c})$ branches soften with fluence, most significantly at zone center ($H = 4$), and resembles the softening with temperature across the Pnma-Cmcm transition. In Fig. 6.3, I show the fluence-dependent and temperature-dependent phonon dispersions in SnSe. In both increased temperatures and increased fluences,

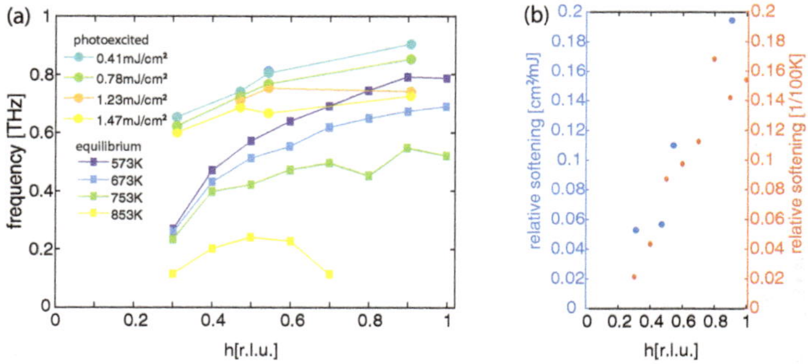

Fig. 6.3 $T(\mathbf{c})$ frequency dependence on fluence and temperature. (a) Photoexcited and thermal equilibrium $T(\mathbf{c})$ phonon dispersions. (b) The relative softening of $T(\mathbf{c})$ phonon, shown for both photoexcitation (defined as relative frequency change per mJ/cm^2) and thermal equilibrium (defined as relative frequency change per Kelvin). Equilibrium data are taken from ref. [22]

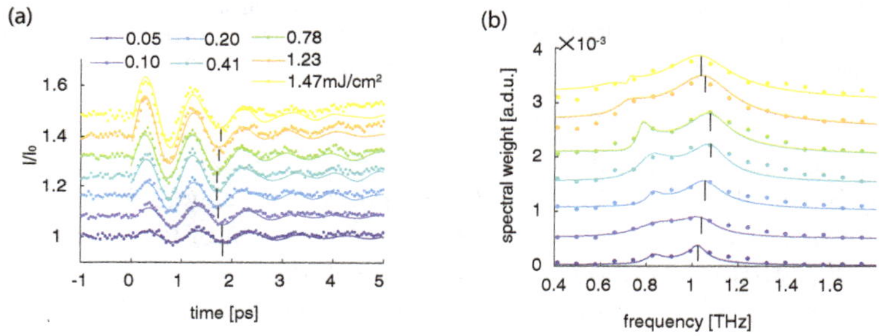

Fig. 6.4 $L(\mathbf{a})$ frequency dependence on fluence. (a) Linear prediction (filled curve) to the integrated scattered intensity (dots) under different fluences. The nonmonotonic fluence dependence of $h = 0.37$ $L(\mathbf{a})$ phonon frequency can be seen from the local minimum marked out by short black lines. (b) Fourier spectrum (dots) of time-domain traces in (a), zooming in near the $L(\mathbf{a})$ phonon frequency. The curves are sum of Lorenzians, each corresponding to a damped harmonic oscillator with parameters extracted from LP. Black short lines mark the LP frequency of $L(\mathbf{a})$ phonons

the phonons soften more toward the folded zone center. See Fig. 6.3(a). And the relative softening (defined as the relative frequency change per mJ/cm^2) for fluence dependence and (defined as the relative frequency change per Kelvin) temperature dependence displays similar behavior up to a scaling. See Fig. 6.3(b). However, this does not imply that the interatomic forces that drive the thermal lattice instability are the same that drive the nonthermal lattice instability. A thermal lattice instability will indicate that as the pump fluence increases, the frequencies of differents change monotonically. However, I show in Fig. 6.4 that $L(\mathbf{a})$ frequency has a significant nonmonotonic fluence dependence. It can be seen in Fig. 6.4(a) that the time stamp of a given phase of I/I_0 shifts backward and then forward in time as the fluence increases. The nonmonotonic fluence dependence can be further confirmed in Fig. 6.4(b), which shows the Fourier spectrum of time domain measurements in Fig. 6.4(a). The $L(\mathbf{a})$ spectral peak of the Lorentzian fits to the damped harmonic

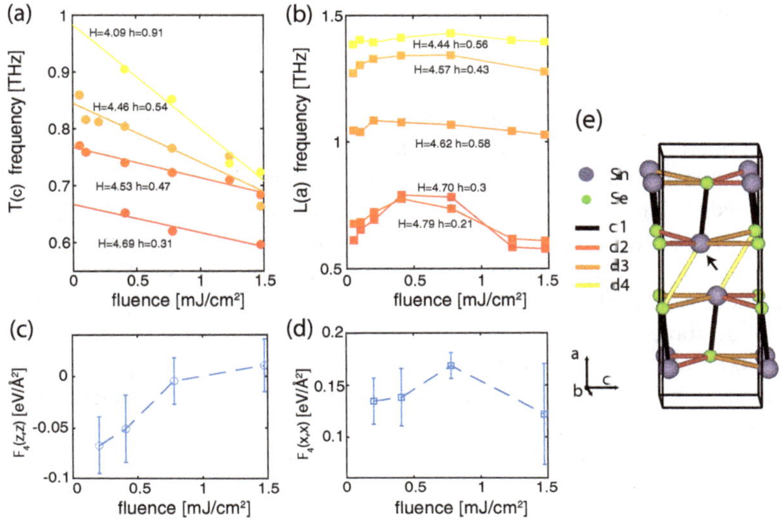

Fig. 6.5 Fluence dependence of photoexcited SnSe phonons and force constants. (a) Nearly linear fluence-dependent softening of $T(\mathbf{c})$ phonon along (H1$\bar{1}$), $H \in [4,5]$ for a subset of **Q** points. (b) Nonmonotonic fluence dependence of $L(\mathbf{a})$ phonon along $(H1\bar{1})$, $H \in [4,5]$ for a subset of **Q** points. (c) The $F_4(z, z)$ forces as fitted from the experimental data. The four data points are an average of fitting results to data sets taken under different $(H1\bar{1})$, $H \in [3,4]$ and $H \in [4,5]$ respectively, but the same selected fluences. The error bars reflect the standard deviation of these fitting results. (d) The $F_4(x, x)$ forces as fitted from the experimental data. (e) The unit cell of the Pnma phase of SnSe illustrating the $d_1 - d_4$ bonds derived from nearest neighbor bonds in the parent cubic structure. The black arrow indicates the space opened up by the d_4 bond tilting away from $\mathbf{a}(x)$ direction

oscillator shifts to higher frequency and then back to lower frequency as the fluence increases. This nonusual nonmonotonic fluence dependence demonstrates the non-thermal nature of pump-induced lattice dynamics.

We summarize in Fig. 6.5(a)–(b) the measured fluence dependence of the $T(\mathbf{c})$ and $L(\mathbf{a})$ modes for multiple H values. The nonmonotonic fluence dependence of the $L(\mathbf{a})$ frequency is most pronounced near the avoided crossing seen in Fig. 6.2. The frequency of $L(\mathbf{a})$ hardens below around 1 mJ/cm^2 and softens as the fluence keeps increasing.

Before we explain the methodology in detail, we show here in Fig. 6.5(c) and (d) the interatomic interactions between the Sn and Se atoms connected by d_4 bond. The interaction strength changes as a function of fluence, revealing the driving mechanism for the lattice instability observed in Chapter 5. Figure 6.5(e) is a reminder of SnSe crystal structure, highlighting the four types of bonds derived from the nearest neighbor bonds in the parent rocksalt structure. From the diffraction results, the change in d_4 bond length and its tilting direction gives strong evidence of a novel lattice instability. The black arrow points to the space opened up by d_4 bond tilting away from $\mathbf{a}(x)$ direction, which is also where the lone pair electrons are localized.

6.2 Nonequilibrium Interatomic Interactions

In order to gain insight into which interatomic interactions are the most responsible for the photoexcited lattice dynamics reflected in the changes of phonon frequencies, we fit an interatomic force model to the measured dispersion.

We assume that the interatomic interactions are renormalized by the photoexcited electron redistribution, without the change in lattice constants in the short probe time window, or significant changes in internal atomic coordinates, which is reported to be on the order of one-thousandth of the lattice constant (see Chap. 5) [15]. To fit the photoexcited phonon dispersions with interatomic forces, we start from the interatomic force constant matrices in real space in the harmonic approximation and allow a limited number of fit parameters to change while fixing others at values calculated from DFT. We define the components of the pair-wise interatomic force tensor between two atoms connected by bond d_n, as $F_n(i, j)$, representing the force on one atom in the ith direction to a unit displacement of the other in the jth direction ($i, j \in x, y, z$).

Each 3×3 matrix $F_n(i, j)$ can be diagonalized to obtain eigenvectors, eigenvalues, and thus phonon dispersions [2]. Note that the diagonalized real-space force matrix $F_n(i, j)$ is not a dynamical matrix. For $F_n(i, j)$ with real eigenvectors, one of the eigenvectors is approximately the vector connecting the two atoms, along the so-called bonding direction, while the other two are approximately perpendicular to the bonding directions. Since the atomic motion is not large enough to change the symmetry of $F_n(i, j)$, we only vary the eigenvalues to minimize the least square errors in the fitting procedure, keeping the eigenvectors fixed. The fitting parameters are the eigenvalue describing force along bonding direction ($f_n(\|)$) and the eigenvalues describing force perpendicular to bonding direction ($f_n(\perp)$). We emphasize that F_n represents the tensor, while f_n represents eigenvalues of the force tensor. For $F_4(i, j)$ that has imaginary eigenvalues (possibly due to the large effect of valence charge that destabilizes the pair-wise ion-ion interaction), the corresponding eigenvectors cannot be related to bonding directions. The fit parameters of a d_4 bond are thus $F_4(x, x)$, $F_4(x, z)$, $F_4(z, x)$, $F_4(z, z)$. Based on the diffraction results under photoexcitation [15], only atomic motions along (a) and (c) axis are observed. We thus assume that the change in force constants associated with atomic displacement along **b**-axis can be neglected.

In Fig. 6.6, we display a total of 17 bonds. d_1–d_{13} are near neighbors chosen in the order of bond lengths. We also consider additional resonant bonds d_{14}–d_{17}. Apart from d_1–d_4 that we will always include in the fitting model, we only select a subset of F_n,($n = 5 - 17$), into the model. We include force constants into the fit in the following order of priority: f_{1-3}($\perp$ and $\|$) and $F_4(i, j)$ ($i, j \neq y$), f_7($\perp$ and $\|$), f_{11} ($\perp$ and $\|$), f_{12} ($\perp$), f_{10} ($\perp$), f_{14} ($\perp$) and f_{15} ($\|$). Bonds apart from d_1–d_4 are sorted in the order of their covariance with the calculated phonon frequency. Including more fit parameters can only further drop the loss function (defined below) by less than 0.1% and shows the sign of overfitting. Note that in the fit, the two $f_n(\perp)$ are constrained to change the scale by the same scaling factor.

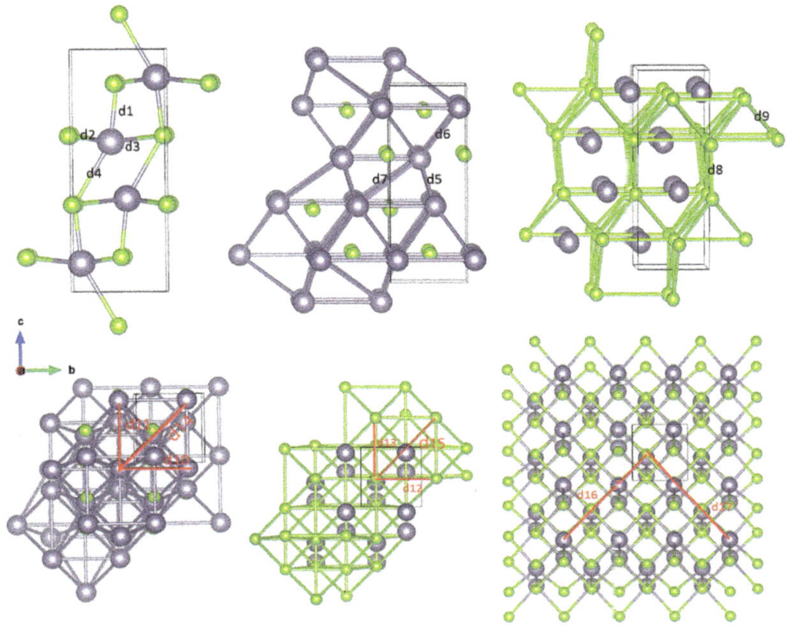

Fig. 6.6 Illustration of candidate bonds $d_1 - d_{17}$

In order to perform the least square fit to the photoexcited phonon dispersion shown in Fig. 6.2, we need to minimize the loss function, which is a sum of weighted squared error $\sum_{\mathbf{q}} W_{\mathbf{q}}(\omega_{\mathbf{q}}^{calc} - \omega_{\mathbf{q}}^{exp})^2$. The weight factor $W_{\mathbf{q}}$ of each data point $(\mathbf{q}, \omega_{\mathbf{q}}^{exp})$ is the signal-to-noise ratio of the time-domain trace sampled at $\mathbf{q}$ (the noise level of the data can be extracted using the well-defined algorithm developed in ref [17, 18]).

It is expected that a robust correlation between the fluence-dependent fitted force constants and the fluence-dependent phonon frequencies is manifested in various fit settings. In Fig. 6.7, we show the fit results for four force constants, $F_4(x, x)$, $F_4(z, z)$, $F_2(\|)$, $F_3(\|)$, under a series of fluences. Apparently, $F_4(z, z)$ shows the same monotonic trend and $F_4(x, x)$ shows the same nonmonotonic trend for different fit settings, and the trends are similar to the fluence dependence of $L(\mathbf{a})$ and $T(\mathbf{c})$, respectively. Below we evaluate the correlation between fitted force constants and observed phonon frequencies. We define target function Y (for either $L(\mathbf{a})$ or $T(\mathbf{c})$) that considers the phonon frequencies of $\mathbf{q}$ points across Γ-X-Γ instead of individual $\mathbf{q}$ points. The target function for $T(\mathbf{c})$ is the sum of product of frequency change and the first reduced Miller index h value of $\mathbf{q}$ points $\mathrm{Y} = \sum_{\mathbf{q}} \Delta\omega_{\mathbf{q}} \times h_{\mathbf{q}}$, based on the linear relationship between $\Delta\omega_{\mathbf{q}}$ and $h_{\mathbf{q}}$. See Fig. 6.3. The target function for $L(\mathbf{a})$ is defined as the sum of absolute change of $L(\mathbf{a})$ frequency across Γ-X-Γ, $\mathrm{Y}= \sum_{\mathbf{q}} \Delta\omega_{\mathbf{q}}$, based on the nonmonotonic fluence dependence of

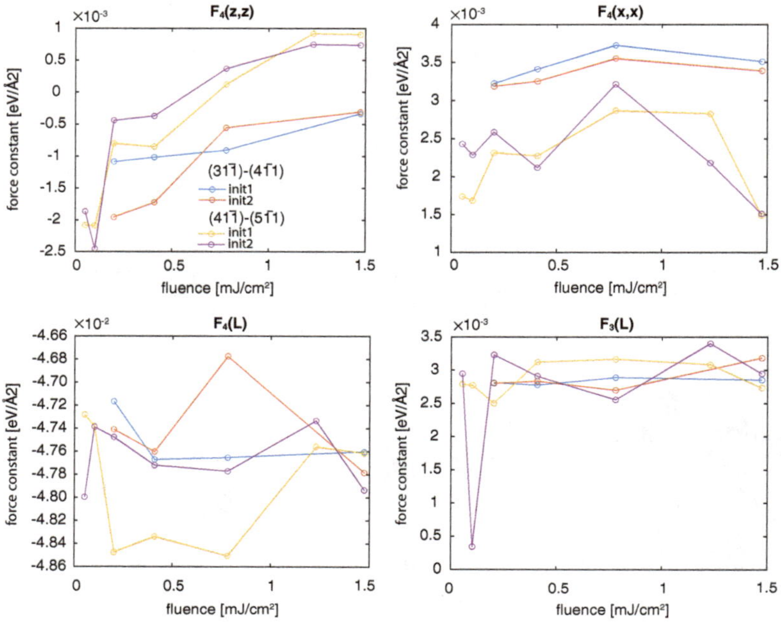

Fig. 6.7 The fitted force constants as a function of fluence. For each force constant, the fit is performed given different reciprocal space regions and different initial fit parameters (init1 and init2)

$L(\mathbf{a})$ across Γ-X-Γ. We will compare, under different settings, r_i, the normalized correlation between X_i (fitted force constant, one of $f_{1-3}(\perp$ and $\parallel)$ and $F_4(i, j)$ $(i, j \neq y)$, $f_7(\perp$ and $\parallel)$, f_{11} $(\perp$ and $\parallel)$, f_{12} $(\perp)$, f_{10} $(\perp)$, f_{14} $(\perp)$ and f_{15} $(\parallel)$) and Y (target function related to the observed phonon frequency). When r_i is nearly 1 or -1, the linear correlation is strong, but when it is nearly 0, the linear correlation is weak. By definition, positive r_i means a positive correlation, and negative r_i a negative correlation. We also compare the linear coefficient k_i assuming linear relation between Y and X_i, under different fit settings. A robust correlation between X_i and Y would imply that under various settings (different initial fit parameters in the least square fitting, different data sets taken with different Brillouin zone cuts, different selections of fit parameters that contain X_i), the $|r_i|$ is consistently large, r_i in all these cases have the same sign, and $k_i r_i$ is large.

We perform the systematic analysis of $T(\mathbf{c})$ target function (Y) and bond parameters (X_i), for data taken on different parts of the reciprocal space; see Fig. 6.8. We further show the systematic analysis of $T(\mathbf{c})$ given different initial least square fit parameters, as well as different fit parameter selections; see Fig. 6.9. From these plots, we conclude that $F_4(z, z)$ best explains the $T(\mathbf{c})$ frequency, because both the quantities $|r_i|$ and $k_i r_i$ are consistently large.

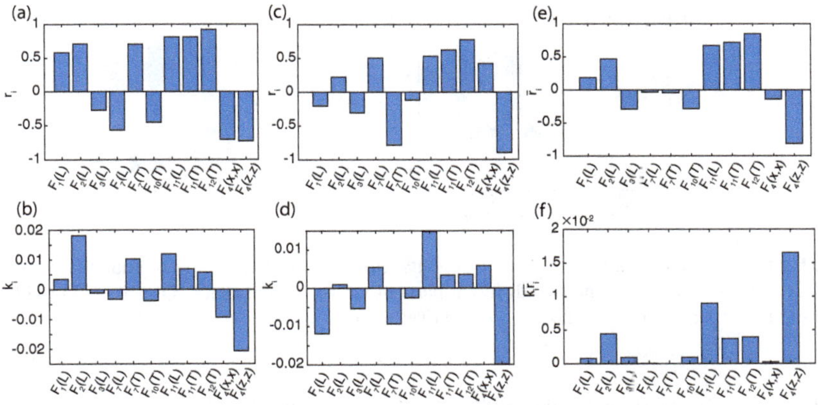

Fig. 6.8 Systematic analysis of $T(\mathbf{c})$ fits to measurements on different Brillouin zones. (a) r_i for the fit to data set taken along $(31\bar{1})$-$(41\bar{1})$. r_i is the correlation between Y and X_i. (b) k_i for the fit to data set taken along $(31\bar{1})$-$(41\bar{1})$. k_i is the linear slope assuming a linear function of Y versus X_i. (c) r_i for the fit to data set taken along $(41\bar{1})$-$(51\bar{1})$. (d) k_i for the fit to data set taken along $(31\bar{1})$-$(41\bar{1})$. (e) Average of (a) and (c), denoted as $\bar{r}_i$. (f) Average $\bar{k}_i$ of (b) and (d), multiplied by $\bar{r}_i$

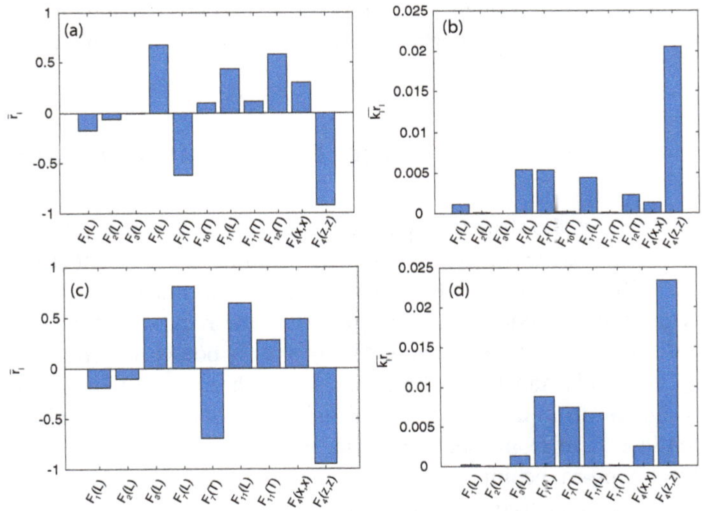

Fig. 6.9 Systematic analysis of $T(\mathbf{c})$ fits given different initial fit parameters and using different force constant subsets. (a) $\bar{r}_i$ averaging r_i of two fits using different initial fit parameters. (b) $\bar{k}_i\bar{r}_i$ multiplying the average quantities $\bar{k}_i$ and $\bar{r}_i$ between two fits using different initial fit parameters. (c) $\bar{r}_i$ averaging r_i of two fits that include both fit parameters as shown on the x-axis. (d) $\bar{k}_i\bar{r}_i$ multiplying the average quantities $\bar{k}_i$ and $\bar{r}_i$ between two fits that include both fit parameters as shown on the x-axis

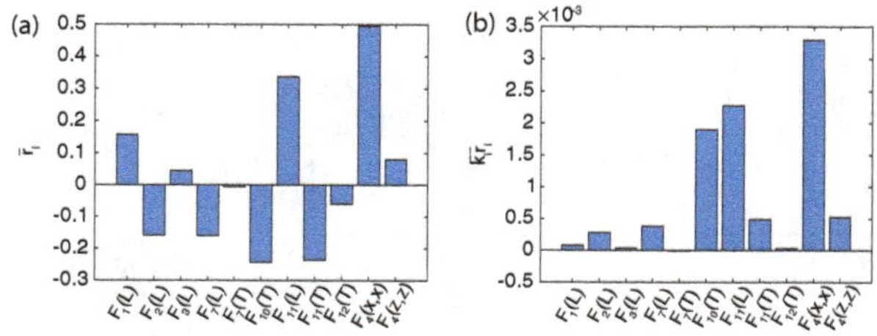

Fig. 6.10 Systematic analysis of $L(\mathbf{a})$ fit results. (a) $\bar{r}_i$ the average of r_i of four fits under two initial conditions of the fit, and two different parts of the reciprocal space, **Q** path (31$\bar{1}$)-(41$\bar{1}$) and (41$\bar{1}$)-(51$\bar{1}$). (b) $\bar{k}_i\bar{r}_i$ where both $\bar{k}_i$ and $\bar{r}_i$ are over the four fits

In Fig. 6.10, we do a similar analysis for $L(\mathbf{a})$ phonon and conclude the $F_4(x, x)$ best explains the $L(\mathbf{a})$ frequency.

In conclusion, we find from our fitting results that a modification to a single force constant $F_4(x, x)$ dominates the observed changes in $L(\mathbf{a})$ frequencies and similarly $F_4(z, z)$ for $T(\mathbf{c})$. Figure 6.5(c)[(d)] shows the average value of $F_4(z, z)$ ($F_4(x, x)$) independently fit to the two data sets $H \in (3,4)$ and $H \in (4,5)$ in Fig. 6.2. Figure 6.5(c) suggests an increase of the initially negative $F_4(z, z)$ upon increased fluence, which is well correlated to the $T(\mathbf{c})$ softening under photoexcitation. The nonmonotonic fluence dependence of fitting results $F_4(x, x)$ as shown in Fig. 6.5(d) is well correlated to the $L(\mathbf{a})$ fluence dependence behavior under photoexcitation.

6.3 A Nonthermal Driving Force for the Lattice Instability

At room temperature, the structure of SnSe belongs to the orthorhombic space group Pnma, which can be understood as a distortion from its cubic parent structure (rock salt). Figure 6.5(e) illustrates the four inequivalent bonds marked in different colors that arise from the six nearest neighbor bonds of the rock salt structure. Notably, the resonant bonding network includes the d_2 and d_3 bonds, which predominantly align along the [011] and [01$\bar{1}$] directions in the **b**–**c** (y–z) plane. These bonds primarily originate from unsaturated p orbitals and possess distinct characteristics that differentiate them from metallic, ionic, or covalent bonds [23]. The resonant bonds exhibit long-range interatomic interactions, significant polarizability, and large Born effective charges. Frozen phonon calculations have revealed that in equilibrium, a soft phonon displacement induces long-range charge-density perturbations along the resonant bonding directions [19] (see Chap. 3). This phenomenon explains the soft phonon behavior observed during the thermal phase transition to a Cmcm structure [19, 21, 24], as well as the substantial anharmonicity that leads to low thermal conductivity [25]. The softening of low-lying transverse optical (TO) phonons in thermal equilibrium is primarily governed by the resonant bonds, particularly the largely in-plane d_2 and d_3 bonds, rather than the out-of-plane

d_4 bond, which does not belong to the resonant bonding network. Interestingly, the most strongly affected Raman active modes during the thermal Pnma-Cmcm transition are those polarized along the in-plane **c** direction, rather than the out-of-plane **a** direction [21, 26].

However, despite including resonant bonds such as d_2 and d_3 in the fitting model, there is a lack of significant correlation between the fitted force tensors for these bonds and the mode softening observed. Conversely, there is a notable correlation between the fluence dependence of the d_4 bond force constants and low-frequency phonons. The reduction in inter-bilayer coupling weakens the frequency of the $T(\mathbf{c})$ mode propagating along the $(h00)$ direction, leading to destabilization of the structure. Based on the fitting results, it is suggested that, in the photoexcited state, the resonant bonds play a lesser role in phonon softening and lattice instability compared to the bilayer-connecting d_4 bond.

We deduce that the interatomic interactions between atoms connected by the d_4 bond are modified by changes in the occupation of the lone pair orbital, which is a mixed orbital consisting of the cation's s orbital and the out-of-plane chalcogen's p orbital. This modification occurs due to the significant spatial overlap between the lone pair orbital and the d_4 bond [19]. In materials with a rock salt parent structure and resonant bonds, the lone pair orbital can have a substantial impact on structural distortions [27–29]. Specifically, in the case of SnSe, the presence of the lone pair orbital causes the breaking of the parent cubic structure into bilayers and leads to the tilting of the d_4 bond away from the **a**-axis direction in the Pnma phase. The lone pair orbital is located in the space created by the tilted bond, adjacent to the d_4 bond (indicated by the black arrow in Fig. 6.1(a)). These findings align with our previous diffraction studies in Chap. 5 [15], where we observed that under photoexcitation, the bilayer-connecting d_4 bond tilts away from the high-temperature Cmcm structure towards a new Immm structure and that the d_4 bond experiences the most significant change in bond length. Thus, the time-resolved diffuse scattering measurements of photoexcited SnSe, presented in this study, suggest that the nonthermal lattice instability at the bonding level is directly caused by the depopulation of the lone pair orbitals and the weakened inter-bilayer coupling. The unconventional nonmonotonic fluence dependence of $F_4(x, x)$ is likely related to the specific details of the excitation density-dependent spatial and temporal relaxation dynamics of the photoexcited carriers. Similar bottleneck effects have been observed in other semiconductors [30–32].

Consistent with the nonthermal behavior, we show below that the photoexcited lattice is more harmonic than the thermal one at temperature that produces similar level of softening of the symmetric Raman active A_g modes at the zone center. We compare the damping versus the frequency change since the latter can be taken as a measure of the proximity to the phase transition [33]. We define the damping rate γ that describes the trajectory of a damped harmonic oscillator (DHO) $x = Ae^{-\gamma t}\cos(\omega_0 t + \phi)$ in the time domain. The response function of the DHO measured in Raman spectroscopy is $\mathrm{Im}\chi(\omega) = \frac{2\omega\gamma}{(\omega_0^2-\omega^2)^2+4\gamma^2\omega^2}$ on the frequency domain, for which the full width at half maximum is 2γ. In Fig. 6.11(a)–

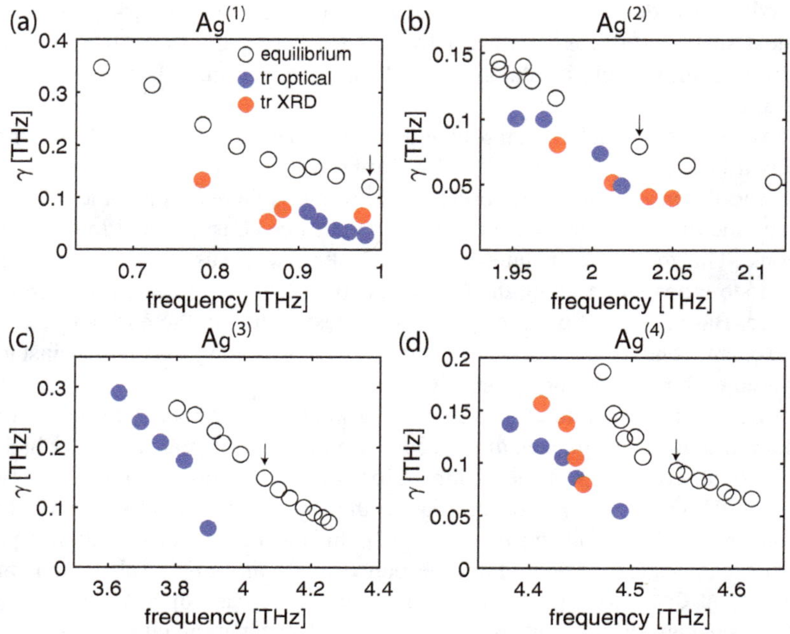

Fig. 6.11 Damping constants (γ) of the phonons compared between the photoexcited states and the thermal equilibrium phase. γ of photoexcited A_g Raman modes [(a) ($A_g^{(1)}$), (b) ($A_g^{(2)}$), (c) ($A_g^{(3)}$), (d) ($A_g^{(4)}$)] are obtained from either the inverse lifetime of time-resolved X-ray scattering measurement (red dots) or time-resolved optical reflectivity (blue dots). Phonon linewidths of photoexcited states are consistently lower than the thermal equilibrium (black circle, measured by Raman scattering [21, 26, 34]) at the same given phonon frequencies. Black arrows correspond to the room temperature Raman measurements

(d), we show γ for all four A_g modes observed under photoexcitation (red dots are results from time-resolved XRD measurements and blue dots are from time-resolved optical reflectivity measurements) and measured by Raman spectroscopy in equilibrium (black open circles). For the same amount of softening, the phonon damping rate (γ) under photoexcitation is consistently lower than its counterpart under thermal equilibrium. The results are particularly notable given that the photoexcited carriers contribute to increasing the phonon linewidth due to the intraband phonon scattering [35–38]. Thus, the observations in Fig. 6.11 indicate that the photoexcited lattice is more harmonic than the lattice at similar proximity (using the frequency softening as a metric) to the thermal phase Cmcm, which is consistent with our finding that photoinduced instability is not mainly originated from resonant bonding network that gives rise to the large anharmonicity, but rather, the bilayer-connecting d_4 bond. Such observations have to do with the fact that anharmonicity of a solid in thermal equilibrium is developed through a significant change in the lattice constants and internal atomic coordinates, which do not happen under photoexcitation on short timescales.

X-ray scattering in the photoexcited states enables us to establish the connection among uniform ($\mathbf{q} = 0$) structural distortions, the atomic bonding inferred from the transient excited-state phonon dispersion ($\mathbf{q} \neq 0$), and the perturbed electron orbitals in the photoexcited states. Previously, only partial indications of such a relationship were possible through measurements of macroscopic properties related to the electron response function (e.g., Born effective charge, optical constants) at $\mathbf{q} = 0$ in the photoexcited states, or microscopic properties at $\mathbf{q} \neq 0$ in the equilibrium states.

In the case of SnSe, we have identified that the photoexcited lattice instability is primarily driven by changes in the bonds connecting the bilayers, which are caused by nonequilibrium carrier distribution. This contrasts with thermal equilibrium, where soft and anharmonic phonons are produced through long-range charge perturbations in resonant bonds. These findings have significant implications for ferroelectrics, where electron redistribution strongly influences lattice dynamics, as well as for phase change materials, where structural changes might be coupled with changes in bonding types.

The methodology we have demonstrated can directly unveil the interplay between electron orbitals, atomic bonds, and structural instability. Consequently, it can be utilized to rationalize approaches for controlling materials and designing novel functional properties in both equilibrium and nonequilibrium states.

Appendix

Oscillator Parameter Extraction for Massive Data

To extract the frequencies of the oscillators, we face the choice of fitting either the time-domain data or its Fourier transform. However, fitting the frequency-domain data using least square methods requires meticulous adjustment of initial fit parameters for each independent time trace. Given the substantial amount of data generated across the reciprocal space, this approach is impractical.

To address this challenge, we opt to perform fitting in the time domain using a singular value decomposition (SVD)-based linear prediction algorithm [16]. For each $\mathbf{q}$ point, the data is shaped into a T $\times$ D matrix, where each row represents a small time window of the data. The linear prediction algorithm assumes that the data can be expressed as a sum of damped harmonic oscillators with amplitude, frequency, decay constants, and phase.

In an ideal scenario without noise, the oscillator parameters can be deterministically retrieved by decomposing the data into several singular value components and solving nonlinear equations [16]. However, in the presence of noise, the SVD approach can be augmented with noise filtering and the retrieval of noise-corrupted modes [17, 18]. We incorporate this approach into the linear prediction algorithm.

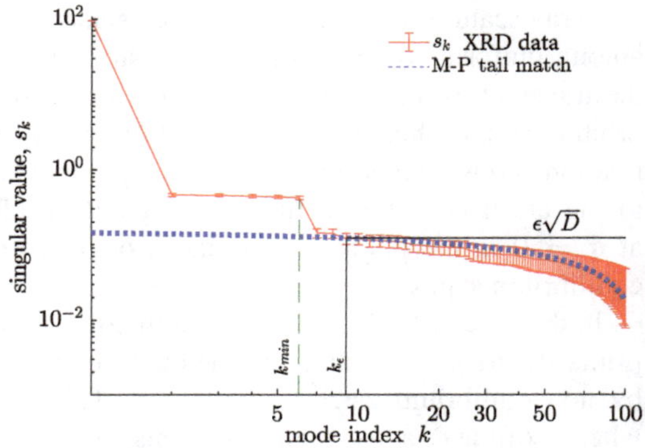

Fig. 6.12 The singular value distribution. The k_{min} and k_ϵ are different types of cutoff mode indices. The error bars are the 95% confidence interval of the singular values. The dashed blue curve corresponds to the scaled Marchenko-Pastur distribution

The noise level of the measurement can be estimated by fitting the tail of the singular value distribution, which represents the singular value as a function of mode index [17, 18]. The fit assumes a scaled Marchenko-Pastur (M-P) distribution, which characterizes the distribution of singular values for a random matrix with uniform errors in its entries. In a system with a limited number of oscillators, the signal is primarily captured in the first few SVD components. By fitting the tail of the singular value distribution, we can not only obtain the intrinsic parameter ϵ of the M-P distribution, which describes the noise level of the data [17], but also determine an appropriate SVD index beyond which the high-index components can be discarded to filter out noise from the data, see Fig. 6.12 [17].

The choice of the cutoff mode index, denoted as k_{min}, serves the dual purpose of noise filtering and minimizing the loss in mode reconstruction. An optimal k_{min} is selected to strike a balance between noise removal (requiring the cutoff index not to be excessively large) and minimizing the loss in mode representation (requiring the cutoff index not to be excessively small, as that would result in an incomplete set of modes to describe the data).

This enhanced and automated version of linear prediction mitigates the introduction of unpredictable systematic errors stemming from human biases in selecting the singular value cutoff index.

Phonon Dispersion and Branch Sorting

In the phonon polarization vector $\epsilon_{\mathbf{q},j}$, it is the index j that brings the concept of the phonon branch. The eigenvalues can be sorted and assigned to specific phonon branches. However, when different branches cross, the concept of "phonon branch" does not practically hold if one just looks at the eigenvalue frequencies at a series of **q** points. Therefore, we need to take into account the continuity of eigenvectors to sort the branches.

Branch sorting based on eigenvector projection between adjacent $\mathbf{q}$ points along a 1D path can be recast into a "distance" matrix trace optimization problem. For $\mathbf{q}$ indexed n and its neighboring $\mathbf{q}$ indexed n+1, the "distance" between a mode $\omega_n^{(i)}$ with eigenvector $\xi_n^{(i)}$ at $\mathbf{q}_n$ and a mode $\omega_{n+1}^{(j)}$ with eigenvector $\xi_{n+1}^{(j)}$ at $\mathbf{q}_{n+1}$ can be defined as $\frac{\omega_{n+1}^{(j)}-\omega_n^{(i)}}{1-\xi_n^{(i)}\cdot\xi_{n+1}^{(j)}}$. Here i and j are the branch indices we want to sort. The definition of "distance" is based on the fact that a candidate mode j which belongs to the same branch as i will have a small difference in frequency as well as a large overlap in eigenvector. One then needs to minimize the trace of this [24×24] "distance" matrix by reshuffling its rows and columns, so that the "distance" sum is the minimum. This problem can be solved in polynomial time, using the Hungarian algorithm [39]. Phonon branch sorting allows the least square fitting to $T(\mathbf{c})$ and $L(\mathbf{a})$ phonons, which switch the order as they disperse (see Fig. 6.2).

Coherent Phonons in SnSe Excited Along $\Gamma - X$

We show that we excite coherent phonons along $\Gamma - X$ below.

We exclusively observe phonon coherences along the reduced wave vector $\mathbf{q} = \mathbf{Q} - \mathbf{G}_{hkl} = (h00)$. In Fig. 6.13, we present two sets of figures where we overlay DFT calculations onto the 2D Fourier spectrum of time-domain diffuse scattering.

In the first set, shown in Fig.6.13(a), we assume that the DFT dispersions extend along the Γ-X-Γ direction. In the second set, depicted in Fig. 6.13(b), we utilize $\mathbf{q}$ values mapped from the sample orientation and detector placement. Generally, we anticipate that the latter set, which takes into account the experimental conditions, would better match the measured phonon dispersions. However, we find that Fig. 6.13(a) provides a significantly better agreement than Fig. 6.13(b), particularly at low $\mathbf{q}$ values near the Brillouin zone center.

It is important to note that this discrepancy is not due to an erroneous $\mathbf{q}$ mapping. During the measurement of diffuse scattering, we do not optimize for the Bragg condition, resulting in the Ewald sphere not intersecting the reciprocal lattice points. As a consequence, the finite curvature of the Ewald sphere leads to a deviation from a straight line connecting the zone centers Γ along $(H1\bar{1})$ in three-dimensional space. However, despite this expected deviation, we still observe a good agreement between the experimental data and simulations assuming a perfect $\mathbf{q}$ path along Γ-X-Γ across multiple zones.

Based on these findings, we propose that the phonon excitation occurs exclusively along the Γ-X direction, with a finite distribution of wave vectors perpendicular to Γ-X. Even when we sample $\mathbf{q}$ points that do not strictly lie on Γ-X, we still measure phonon excitations along the Γ-X direction.

The consistency between the observed frequencies of excitations and the calculated phonon frequencies suggests that the non-zone center excitations are coherent phonons, rather than squeezed phonons [11, 13] that come with overtones.

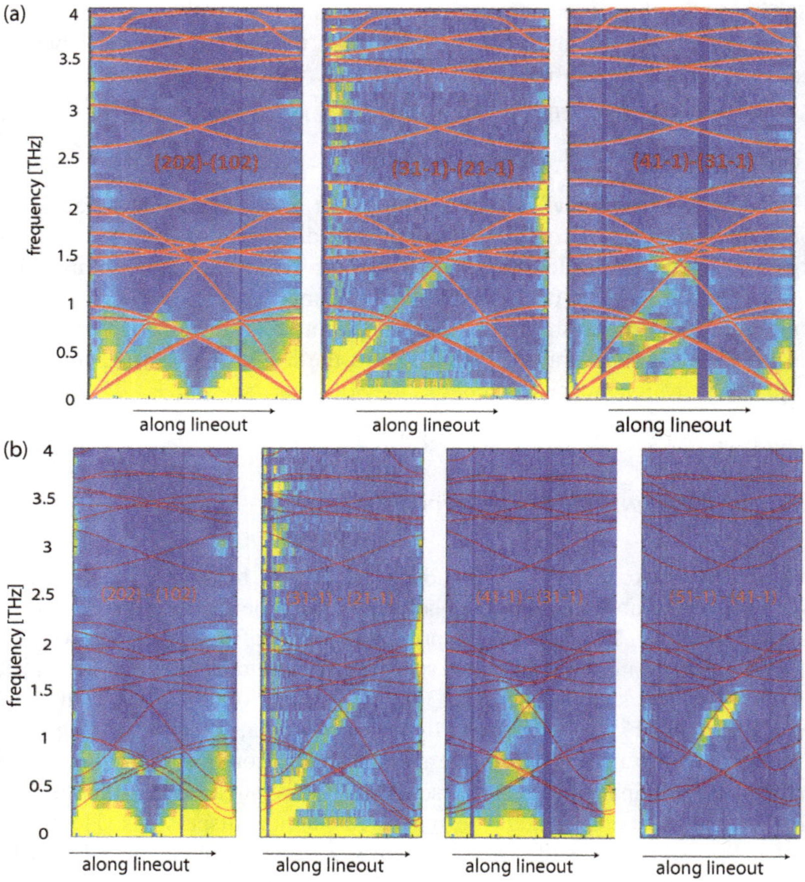

Fig. 6.13 (a) The data overlaid with DFT ground-state dispersion, which assumes that the **q** points are strictly along Γ-X-Γ. (b) The data overlaid with DFT ground-state dispersion, which uses the **q** mapped from calibrated sample orientation and detector placement

To further confirm the coherent phonon character, we examine the scattering intensity of longitudinal and transverse phonons simultaneously. In Fig. 6.14(c), we plot the Fourier spectrum of the scattering intensities at different zone boundaries X. It is evident that the scattering intensity of the two X phonons exhibits different **Q** dependence.

By comparing $\mathbf{Q} = (2.5, 1, \bar{1})$, $\mathbf{Q} = (3.5, 1, \bar{1})$, and $\mathbf{Q} = (4.5, 1, \bar{1})$, we observe that the higher-frequency mode (around 1.35 THz) significantly increases its intensity as the H index increases, which is expected for an **a**-polarized phonon. This behavior is in sharp contrast to the low response of the lower-frequency mode (around 0.75 THz) to the variation of H. Such observations provide evidence that

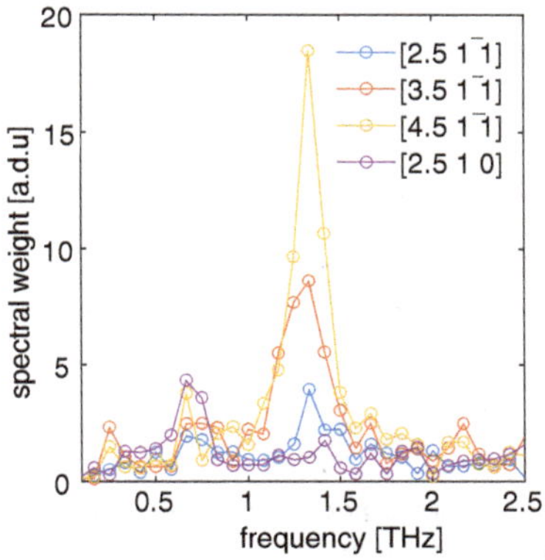

Fig. 6.14 The Fourier spectrum of the normalized scattering intensity for several **Q**-points at Brillouin zone boundary X (reduced Miller index h takes half odd integer values). Fluence is 0.78 mJ/cm^2. The two low-frequency acoustic phonon branches can be identified as $T(\mathbf{c})$ (the lower-frequency branch) and $L(\mathbf{a})$ (the higher-frequency branch) by comparing the scattering intensities measured on different parts of the reciprocal space

the higher-frequency phonon is polarized along the **a** crystal axis ($L(\mathbf{a})$), while the lower-frequency phonon is mostly polarized orthogonal to the **a** axis.

The fact that the coherence at $\mathbf{Q} = (1.5, 0, 2)$ produces a higher spectral weight of the lower-frequency phonon than **Q** points with smaller absolute l values, and the fact that the spectral weight does not disappear at $k = 0$ (which would be the case if the phonon were only polarized along the b axis), indicate that the lower-frequency phonon is a transverse acoustic phonon polarized along c, i.e., $T(\mathbf{c})$. The observed features of an avoided crossing between $T(\mathbf{c})$ and $L(\mathbf{a})$ are consistent with the DFT calculations.

All the evidence presented points to the fact that we are dealing with a one-dimensional coherent excitation. When fitting the phonon dispersions, we can ignore the deviation of the k and l indices from integer values.

These coherent phonons along $\mathbf{q} = (h00)$ may be generated through local disorders that provide large wave vector momentum matching, and they are detected through a heterodyning mechanism [40].

We analyze various properties of the excitation to gain insights into its coherent nature. We observe that the modulation amplitude of the intensity for both $L(\mathbf{a})$ and $T(\mathbf{c})$ modes shows a linear relationship with the fluence. Since we are dealing with coherent phonons and the fluence is directly proportional to the phonon amplitude, we can deduce that the phonon amplitude scales proportionally with the laser fluence. This observation aligns with the characteristics of a coherent Raman excitation, where the phonon amplitude is expected to be directly proportional to the intensity of the incident light field [41, 42].

In the Raman process, electron-phonon coupling plays a crucial role. In Chap. 2, I focused solely on the excitation of phonons at the zone center, as the photons'

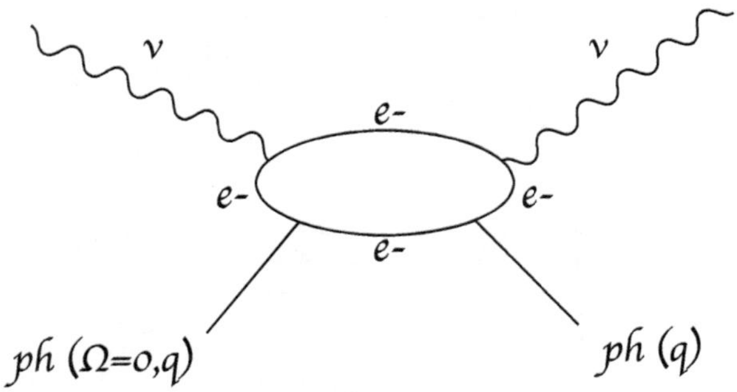

Fig. 6.15 Interaction diagram that describes coherent phonon excitation at $\mathbf{q} \neq 0$. ν represents a photon and $e-$ represents an electron. ph represents a phonon, $ph(\Omega = 0, \mathbf{q})$ is a static field that is most likely associated with the lattice distortion, $ph(\mathbf{q})$ is the experimentally observed coherent acoustic phonon. ν represents photon and e represents electron

wave vector is significantly smaller than the Brillouin zone, allowing momentum matching only for zone center phonons. However, to explain our experimental findings, we need to consider additional interaction terms involving another field with a high wave vector. We have already ruled out the possibility of squeezed phonon pair excitation. Therefore, it is likely that the field providing the momentum matching is a static field with a large wave vector [40], which can be interpreted as phonon fields with zero frequency ($\Omega = 0$). The interaction diagram I have in mind is depicted in Fig. 6.15.

I briefly touched on the interaction $e - ph(\mathbf{q})$ in Chap. 2. There are, in general, two types of electron-acoustic phonon coupling, one based on long-range Coulomb interaction and the other one based on short-range deformation potential. Assuming long wavelength limit, the short-range interaction involves only longitudinal acoustic phonon, via a term $H_{e,\mathrm{L}} \propto (u\mathbf{q}\boldsymbol{\epsilon})$ where $u\boldsymbol{\epsilon}$ is the phonon field (with amplitude u and polarization $\boldsymbol{\epsilon}$) and $\mathbf{q}$ is the wave vector of the phonon [43]. For the transverse acoustic phonon, $e - ph(\mathbf{q})$ likely involves piezo-acoustic coupling. It is possible that in SnSe, $L(\mathbf{a})$ and $T(\mathbf{c})$ are excited through different types of coupling—though involving the same field of distortion.

To gain further insights, we examine the initial phases of the coherent acoustic phonons. In Figs. 6.16 and 6.17, we present the initial phases of $L(\mathbf{a})$ and $T(\mathbf{c})$ phonons at multiple $\mathbf{q}$ points along the Γ-X-Γ direction.

For the $L(\mathbf{a})$ phonon, we observe that the oscillation starts resembling a sine function or a cosine function with a phase of $-\frac{\pi}{2}$. This behavior is illustrated in Fig. 6.16(a). The consistency of the $L(\mathbf{a})$ phonon starting like a sine function throughout the probed reciprocal space along Γ-X-Γ can be verified at various fluences, as shown in Fig. 6.16(b).

On the other hand, Fig. 6.17(a) reveals that the $T(\mathbf{c})$ phonon exhibits an initial oscillation resembling a cosine function, with a phase of either 0 or π. Notably, the $T(\mathbf{c})$ phonon displays different initial phases at low and high $\mathbf{q}$ values along the Γ-X-Γ path. Specifically, at low $\mathbf{q}$ ($h < 0.5$), the phase is 0, while at high $\mathbf{q}$

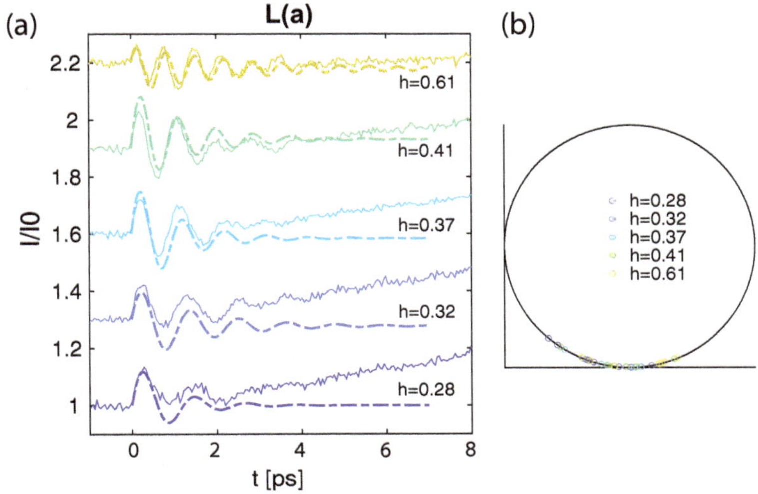

Fig. 6.16 (a) $L(\mathbf{a})$ phonon as a harmonic oscillator starting at phase $-\frac{\pi}{2}$, or starting like a sine function. At the five selected **q** points, the time-resolved diffuse scattering intensity (solid lines) together with its $L(\mathbf{a})$ phonon component (dashed lines) is shown. Data displayed is taken under 1.47 mJ/cm^2. (b) Of the data sets at different fluences and **q** points, the phases of $L(\mathbf{a})$ component are shown on a unit radius circle. Fluences for the same **q** point are not specified in the plot

($h > 0.5$), the phase is π. This phase difference holds true for various probing fluences, as depicted in Fig. 6.17(b). Moreover, the $T(\mathbf{c})$ phonon very close to the X point appears to exhibit an initial phase of $\frac{\pi}{2}$, which lies halfway between 0 and π, as illustrated in Fig. 6.17(c) and (d).

These observations suggest that the driving force for the $L(\mathbf{a})$ phonon is short-lived compared to the phonon frequency, while the driving force for the $T(\mathbf{c})$ phonon persists for a longer duration.

The coherent excitations of acoustic phonons along Γ-X-Γ in SnSe are attributed to defects present in the material. According to ref. [44], a significant density of point defects is known to form in SnSe during crystal growth processes, including the Bridgman technique employed in the growth of our SnSe samples. These defects lead to a local arrangement where two second-nearest-neighboring SnSe monolayers with the same dipole orientation are interconnected, forming a stacking fault sequence. Importantly, these stacking faults do not disrupt the crystal structure in the **b–c** plane and result in a defect density Fourier distribution primarily confined along the Γ-X-Γ direction.

In Fig. 6.18, we present evidence supporting the presence of defects in SnSe. The red curve represents the diffuse scattering intensity along a white linecut in Fig. 6.18(a), which approximately follows the path of Γ-X-Γ. The blue curve represents the simulated thermal diffuse scattering intensity. The additional broad diffuse scattering observed near the X point, superimposed on the thermal phonon scattering, provides evidence of static disorder induced by defects. A disorder which

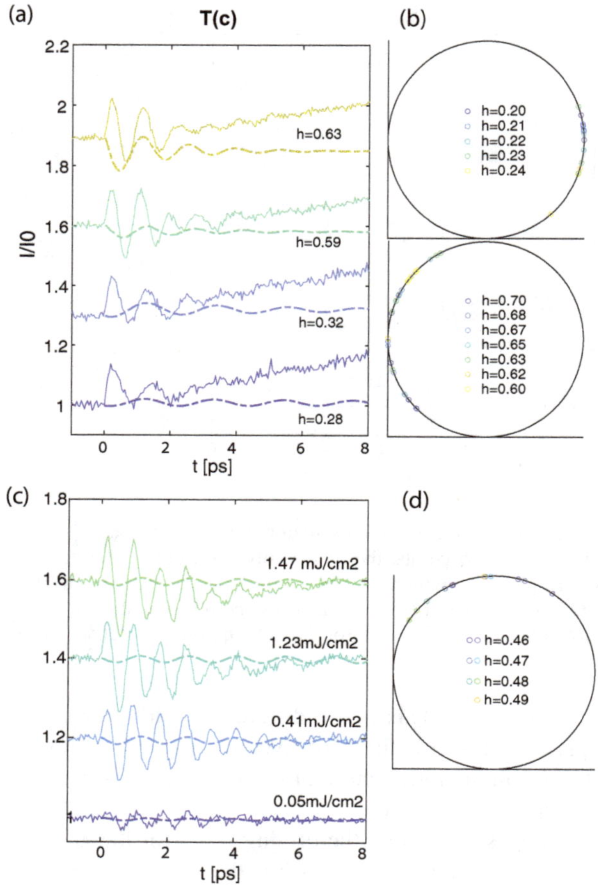

Fig. 6.17 (a) $T(\mathbf{c})$ phonon as a harmonic oscillator starting at phase π at high $\mathbf{q}$ (h >0.5) or 0 at low $\mathbf{q}$ ($h < 0.5$). Data displayed is taken under 1.47 mJ/cm^2. (b) Of the data sets at different fluences and $\mathbf{q}$ points, the phases of the $T(\mathbf{c})$ component are shown on a unit radius circle. Fluences for the same $\mathbf{q}$ point are not specified in the plot. (c) $T(\mathbf{c})$ phonon near X point ($h = 0.47$) as a harmonic oscillator starting at phase $\frac{\pi}{2}$. (d) Statistical initial phase distribution near $h = 0.47$. The sampling of $\mathbf{q}$ point is very dense and thus different $\mathbf{q}$ points may approximate the same h value in two-digit precision. Fluences for the same $\mathbf{q}$ point are not specified in the plot

appears around X point is not inconsistent with the fact that coherent $T(\mathbf{c})$ flips its sign when $\mathbf{q}$ goes from one side of X to the other. The nature of the disorder that is relatively localized near X is not clear.

Further investigation into defect chemistry is required to fully understand this phenomenon. However, such an analysis is beyond the scope of the current study, which focuses on the coherent excitation of acoustic phonons and its implications.

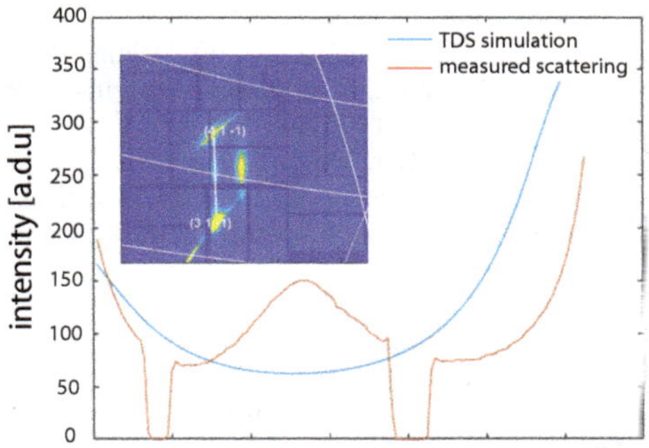

Fig. 6.18 Static diffuse scattering of SnSe. The red curve shows static (pump laser off) diffuse scattering intensity (offimg) along the white linecut in inset from (41-1) to (31-1), approximately the path of Γ-X-Γ. The blue curve is the thermal diffuse scattering (TDS) intensity simulated using the DFT interatomic forces, scaled to match the near-zone-center intensity level as measured

References

1. Y. Huang, S. Teitelbaum, S. Yang, G. De la Peña, T. Sato, M. Chollet, D. Zhu, J.L. Niedziela, D. Bansal, A.F. May, A.M. Lindenberg, O. Delaire, M. Trigo, D.A. Reis, Nonthermal bonding origin of a novel photoexcited lattice instability in SnSe. Phys. Rev. Lett. **131**(15), 156902 (2023). https://doi.org/10.1103/PhysRevLett.131.156902
2. M. Born, K. Huang, M. Lax, Dynamical theory of crystal lattices. Amer. J. Phys. **23**(7), 474–474 (1955)
3. E.G. Brovnman, Y. Kagan, The phonon spectrum of metals. Sov. Phys. JETP **25**(364), 104 (1967)
4. L.J. Sham, Electronic contribution to lattice dynamics in insulating crystals. Phys. Rev. **188**, 1431–1439 (1969)
5. M. Krisch, F. Sette, *Inelastic X-Ray Scattering from Phonons* (Springer, Berlin, 2007), pp. 317–370
6. A.Q.R. Baron, *High-Resolution Inelastic X-Ray Scattering II: Scattering Theory, Harmonic Phonons, and Calculations* (Springer International Publishing, Cham, 2014), pp. 1–32
7. A.Q.R. Baron, *High-Resolution Inelastic X-Ray Scattering I: Context, Spectrometers, Samples, and Superconductors* (Springer International Publishing, Cham, 2014), pp. 1–68
8. B.N. Brockhouse, A.T. Stewart, Scattering of neutrons by phonons in an aluminum single crystal. Phys. Rev. **100**, 756–757 (1955)
9. R.A. Cowley, Lattice dynamics and phase transitions of strontium titanate. Phys. Rev. **134**, A981–A997 (1964)
10. D.N. Basov, R.D. Averitt, D. Hsieh, Towards properties on demand in quantum materials. Nat. Mater. **16**(11), 1077–1088 (2017)
11. M. Trigo, M. Fuchs, J. Chen, M.P. Jiang, M. Cammarata, S. Fahy, D.M. Fritz, K. Gaffney, S. Ghimire, A. Higginbotham, S.L. Johnson, M.E. Kozina, J. Larsson, H. Lemke, A.M. Lindenberg, G. Ndabashimiye, F. Quirin, K. Sokolowski-Tinten, C. Uher, G. Wang, J.S. Wark, D. Zhu, D.A. Reis, Fourier-transform inelastic X-ray scattering from time- and momentum-dependent phonon-phonon correlations. Nat. Phys. **9**(12), 790–794 (2013)
12. D. Zhu, A. Robert, T. Henighan, H.T. Lemke, M. Chollet, J.M. Glownia, D.A. Reis, M. Trigo, Phonon spectroscopy with sub-mev resolution by femtosecond X-ray diffuse scattering. Phys. Rev. B **92**, 054303 (2015)

13. M.P. Jiang, M. Trigo, I. Savić, S. Fahy, É.D. Murray, C. Bray, J. Clark, T. Henighan, M. Kozina, M. Chollet, J.M. Glownia, M.C. Hoffmann, D. Zhu, O. Delaire, A.F. May, B.C. Sales, A.M. Lindenberg, P. Zalden, T. Sato, R. Merlin, D.A. Reis, The origin of incipient ferroelectricity in lead telluride. Nat. Commun. **7**(1), 12291 (2016)
14. S.W. Teitelbaum, T.C. Henighan, H. Liu, Mason P. Jiang, D. Zhu, M. Chollet, T. Sato, É.D. Murray, S. Fahy, S. O'Mahony, T.P. Bailey, C. Uher, M. Trigo, D.A. Reis, Measurements of nonequilibrium interatomic forces using time-domain X-ray scattering. Phys. Rev. B **103**, L180101 (2021)
15. Y. Huang, S. Yang, S. Teitelbaum, G. De la Peña, T. Sato, M. Chollet, D. Zhu, J.L. Niedziela, D. Bansal, A.F. May, A.M. Lindenberg, O. Delaire, D.A. Reis, M. Trigo, Observation of a novel lattice instability in ultrafast photoexcited SnSe. Phys. Rev. X **12**, 011029 (2022)
16. H. Barkhuijsen, R. de Beer, W.M.M.J. Bovee, D. van Ormondt, Retrieval of frequencies, amplitudes, damping factors, and phases from time-domain signals using a linear least-squares procedure. J. Mag. Reson. (1969) **61**(3), 465–481 (1985)
17. B.P. Epps, E.M. Krivitzky, Singular value decomposition of noisy data: noise filtering. Experim. Fluids **60**(8), 1–23 (2019)
18. B.P. Epps, E.M. Krivitzky, Singular value decomposition of noisy data: mode corruption. Experim. Fluids **60**(8), 1–30 (2019)
19. C.W. Li, J. Hong, A.F. May, D. Bansal, S. Chi, T. Hong, G. Ehlers, O. Delaire, Orbitally driven giant phonon anharmonicity in SnSe. Nat. Phys. **11**(12), 1063 (2015)
20. D. Bansal, J. Hong, C.W. Li, A.F. May, W. Porter, M.Y. Hu, D.L. Abernathy, O. Delaire, Phonon anharmonicity and negative thermal expansion in SnSe. Phys. Rev. B **94**, 054307 (2016)
21. T. Lanigan-Atkins, S. Yang, J.L. Niedziela, D. Bansal, A.F. May, A.A. Puretzky, J.Y.Y. Lin, D.M. Pajerowski, T. Hong, S. Chi, G. Ehlers, O. Delaire, Extended anharmonic collapse of phonon dispersions in SnS and SnSe. Nat. Commun. **11**(1), 1–9 (2020)
22. T. Chatterji, U.D. Wdowik, G. Jagło, S. Rols, F.R. Wagner, Soft-phonon dynamics of the thermoelectric β-SnSe at high temperatures. Phys. Lett. A **382**(29), 1937–1941 (2018)
23. L. Guarneri, S. Jakobs, A. von Hoegen, S. Maier, M. Xu, M. Zhu, S. Wahl, C. Teichrib, Y. Zhou, O. Cojocaru-Miredin, M. Raghuwanshi, C.-F. Schon, M. Drogeler, C. Stampfer, R.P.S.M. Lobo, A. Piarristeguy, A. Pradel, J.-Y. Raty, M. Wuttig, Metavalent bonding in crystalline solids: how does it collapse? Adv. Mater. **33**(39), 2102356 (2021)
24. G. Qin, X. Zhang, S.-Y. Yue, Z. Qin, H. Wang, Y. Han, Hu. Ming, Resonant bonding driven giant phonon anharmonicity and low thermal conductivity of phosphorene. Phys. Rev. B **94**(16), 165445 (2016)
25. S. Lee, K. Esfarjani, T. Luo, J. Zhou, Z. Tian, G. Chen, Resonant bonding leads to low lattice thermal conductivity. Nat. Commun. **5**, 3525 (2014)
26. F. Liu, P. Parajuli, R. Rao, P.C. Wei, A. Karunarathne, S. Bhattacharya, R. Podila, J. He, B. Maruyama, G. Priyadarshan, J.R. Gladden, Y.Y. Chen, A.M. Rao, Phonon anharmonicity in single-crystalline SnSe. Phys. Rev. B **98**(22), 224309 (2018)
27. K.M. Rabe, M. Dawber, C. Lichtensteiger, C.H. Ahn, J.-M. Triscone, Modern physics of ferroelectrics: Essential background, in *Physics of Ferroelectrics* (Springer, Berlin, 2007), pp. 1–30
28. U.V. Waghmare, N.A. Spaldin, H.C. Kandpal, R. Seshadri, First-principles indicators of metallicity and cation off-centricity in the iv–vi rocksalt chalcogenides of divalent Ge, Sn, and Pb. Phys. Rev. B **67**(12), 125111 (2003)
29. L.E. Orgel, The stereochemistry of B subgroup metals. Part II. The inert pair. J. Chem. Soc., 3815–3819 (1959). https://doi.org/10.1039/JR9590003815
30. R.F. Leheny, J. Shah, R.L. Fork, C.V. Shank, A. Migus, Dynamics of hot carrier cooling in photo-excited GaAs. Solid State Commun. **31**(11), 809–813 (1979)
31. H.M. Van Driel, Influence of hot phonons on energy relaxation of high-density carriers in germanium. Phys. Rev. B **19**(11), 5928–5932 (1979)
32. E.J. Yoffa, Screening of hot-carrier relaxation in highly photoexcited semiconductors. Phys. Rev. B **23**, 1909–1919 (1981)

33. M.T. Dove, Theory of displacive phase transitions in minerals. Amer. Mineral. **82**(3–4), 213–244 (1997)
34. W. Peng, F.-R.. Fan, M. Hagihala, M. Kofu, K. Peng, Y. Ishikawa, S. Lee, T. Honda, M. Yonemura, K. Ikeda, T. Otomo, G. Wang, K. Nakajima, Z. Sun, T. Kamiyama, Strong lattice anharmonicity exhibited by the high-energy optical phonons in thermoelectric material. New J. Phys. **22**(8), 083083 (2020)
35. T.A. Fjeldly, F. Cerdeira, M. Cardona, Effects of free carriers on zone-center vibrational modes in heavily doped p-type Si. I. Acoustical modes. Phys. Rev. B **8**(10), 4723–4733 (1973)
36. D. Olego, M. Cardona, Self-energy effects of the optical phonons of heavily doped p-GaAs and p-Ge. Phys. Rev. B **23**(12), 6592–6602 (1981)
37. M. Ledgerwood, H. van Driel, Picosecond phonon dynamics and self-energy effects in highly photoexcited germanium. Phys. Rev. B **54**(7), 4926–4935 (1996)
38. K.T. Tsen, J.G. Kiang, D.K. Ferry, H. Morkoç, Subpicosecond time-resolved Raman studies of LO phonons in GaN: dependence on photoexcited carrier density. Appl. Phys. Lett. **89**(11), 112111 (2006)
39. H.W. Kuhn, The hungarian method for the assignment problem. Naval Res. Logist. Quart. **2**(1–2), 83–97 (1955)
40. M. Kozina, M. Trigo, M. Chollet, J.N. Clark, J.M. Glownia, A.C. Gossard, T. Henighan, M.P. Jiang, H. Lu, A. Majumdar, D. Zhu, D.A. Reis, Heterodyne X-ray diffuse scattering from coherent phonons. Struct. Dyn. **4**, 054305 (2017)
41. R. Merlin, Generating coherent THz phonons with light pulses. Solid State Commun. **102**(2), 207–220 (1997). Highlights in Condensed Matter Physics and Materials Science
42. T.E. Stevens, J. Kuhl, R. Merlin, Coherent phonon generation and the two stimulated raman tensors. Phys. Rev. B **65**, 144304 (2002)
43. M. Cardona, Y.Y. Peter, *Fundamentals of Semiconductors*, vol. 619 (Springer, Berlin, 2005)
44. Z. Wang, C. Fan, Z. Shen, C. Hua, Q. Hu, F. Sheng, Y. Lu, H. Fang, Z. Qiu, J. Lu, Z. Liu, W. Liu, Y. Huang, Z.A. Xu, D.W. Shen, Y. Zheng. Defects controlled hole doping and multivalley transport in SnSe single crystals. Nat. Commun. **9**(1), 47 (2018).

Chapter 7
Quantifying Deformation Potentials of Bi_2Te_3 and Bi_2Se_3

Abstract Quantifying electron-phonon interactions for the surface states of topological materials can provide key insights into surface-state transport, topological superconductivity, and potentially how to manipulate the surface state using a structural degree of freedom. We perform time-resolved X-ray diffraction (XRD) and angle-resolved photoemission (ARPES) measurements on Bi_2Te_3 and Bi_2Se_3, following the excitation of coherent A_{1g} optical phonons. We extract and compare the deformation potentials coupling the surface electronic states to local A_{1g}-like displacements in these two materials using the experimentally determined atomic displacements from XRD and electron band shifts from ARPES. We find the coupling in Bi_2Te_3 and Bi_2Se_3 to be similar and generally in agreement with expectations from density functional theory. We establish a methodology that quantifies the mode-specific electron-phonon coupling experimentally, allowing detailed comparison to theory. Our results shed light on fundamental processes in topological insulators involving electron-phonon coupling.

The ARPES data in this chapter are provided by Jonathan Sobota at Stanford University. The DFT calculations were provided by Jose D. Querales-Flores, Ivana Savic, Stephen Fahy from Tyndall National Institute, and Jiang Cao from ETH Zürich. In this chapter, part of the figures are reprinted from ref [1], with permission from American Physical Society.

7.1 Electron Phonon Lock-in Detection

Electron-phonon interactions are of fundamental importance in condensed matter. In the case of topological insulators, the scattering properties of their surface states (SS) have a direct impact on their application to energy-efficient electronics [2–6]. In this context, there have been studies that extract the electron-phonon coupling

Y. Huang, *Towards the Optical Control of Resonantly Bonded Materials*, Springer Theses, https://doi.org/10.1007/978-3-031-42826-5_7

integrated over phonon modes using ARPES [7–11], helium atom scattering [12, 13, 13], and transport measurements [14]. The metal-intercalated Bi_2Se_3 family of topological insulators are also topological superconductor candidates [15–20], where electron-phonon coupling potentially mediates unconventional pairing symmetries [21]. Therefore, quantifying the mode-specific electron-phonon coupling may benefit the search for and understanding of topological superconductors. Studying electron-phonon coupling is also beneficial for understanding and designing functional topological materials. For example, the topological SS have been predicted to facilitate certain chemical reactions [22, 23], among which some benefit from phonon participation [24]. Finally, electron-phonon interactions hold implications for optical control of matter [25, 26], for example, by ultrafast manipulation of SS transport [27], or by exciting certain lattice degrees of freedom to switch topological phases [28–31].

The deformation potential (DP) quantifies the electron-phonon coupling of a specific electronic band to a specific phonon mode as a change of electronic energy for a given lattice displacement [32]. Ultrafast spectroscopy has demonstrated the unique capability to measure this band- and mode-resolved electron-phonon coupling [33, 34]. Here, we report ultrafast measurements of the band- and mode-specific DPs in the prototypical topological insulators Bi_2Te_3 and Bi_2Se_3. We measure the DPs of the SS coupled to A_{1g} phonons by combining the atomic displacements measured by XRD with the energy shift of the SS measured by time-resolved ARPES. We find that the experimental DP of the $A_{1g}^{(1)}$ phonon mode is 1.4 times larger in Bi_2Te_3 than in Bi_2Se_3. Within experimental uncertainties, the DPs are indistinguishable and on the order of 2 meV/pm. This qualitatively agrees with the naive expectation for two isostructural and isoelectronic materials. Predictions from density functional theory (DFT) match experimental values within a factor of 2.5. DFT correctly predicts a larger DP in Bi_2Te_3 than in Bi_2Se_3. Constrained DFT calculations of the excited state reproduce the experimentally observed mode softening well. The consistent experimental values and their general agreement with theory validate our experimental and theoretical concepts while also pointing to their challenges. Our mode- and band-specific measurement of the DPs in Bi_2Te_3 and Bi_2Se_3 contributes to the understanding of SS transport as electron-phonon coupling is an important contribution to the scattering rates.

There is no universal convention for defining DP involving optical phonons. Here we define the SS DPs with respect to A_{1g} modes in Bi_2Te_3 and Bi_2Se_3:

$$\mathrm{DP} = \frac{\Delta E_{\mathrm{SS}}^{A_{1g}}}{u_{\mathrm{SS}}^{A_{1g}}}, \tag{7.1}$$

where $u_{\mathrm{SS}}^{A_{1g}}$ is an atomic displacement near the surface projected onto the bulk A_{1g} mode eigenvector (as shown in Fig. 7.1(a)) and averaged over the spatial extent of the surface state. This definition enables us to establish a formal surface deformation potential in relation to a bulk photon, despite the potential subtle differences in interatomic forces near the surface compared to those in the bulk region. $\Delta E_{SS}^{A_{1g}}$ represents the energy shift of the surface state (SS) caused by the atomic displacement mentioned earlier. For ease of comparison between the two materials, we choose to quantify $u_{SS}^{A_{1g}}$ with respect to the bismuth displacements. To simplify the notation, we will omit the A_{1g} subscript.

When combining XRD and ARPES data to calculate DPs, it is crucial to consider the distinct characteristics of these methods as bulk- and surface-sensitive probes, respectively. This consideration becomes particularly important due to the excitation of the sample by a laser with a finite penetration depth and subsequent diffusion of the excited carriers, resulting in inhomogeneous phonon displacements within the sample. Such effects can introduce systematic errors in the analysis, necessitating substantial correction based on specific experimental details. We find that the experimental DP of the $A_{1g}^{(1)}$ phonon mode is 1.4 times larger in Bi_2Te_3 than in Bi_2Se_3. Within experimental uncertainties, these DPs are similar and on the order of 2 meV/pm. This qualitatively agrees with the naive expectation for two isostructural and isoelectronic materials. Predictions from density functional theory (DFT) match experimental values within a factor of 2.5. DFT correctly predicts a larger DP in Bi_2Te_3 than in Bi_2Se_3. Constrained DFT calculations of the excited state reproduce the experimentally observed mode softening well. The consistent experimental values and their general agreement with theory validate our experimental and theoretical concepts while also pointing to their challenges.

7.2 Experimental Details

In both XRD and ARPES experiments, the materials are subjected to room temperature conditions and excited using ultrafast 1.55 eV laser pulses. These laser pulses serve to excite fully symmetric coherent optical A_{1g} phonons, which, in turn, lead to modulations in the XRD intensities. By analyzing these modulations, we can determine the amplitudes of phonon motions. The coherent A_{1g} phonons also generate electron band shifts through electron-phonon coupling. This results in oscillations in the energy levels of the electron bands, which we quantify by employing time-resolved ARPES techniques.

We begin by discussing the XRD measurements conducted on Bi_2Te_3. Figure 7.1(a) illustrates the atomic motion diagram, specifically highlighting the two A_{1g} modes of interest. To determine the photoinduced atomic displacements associated with these phonon modes, a time-resolved XRD experiment was performed at the X-ray pump-probe (XPP) instrument of the Linac Coherent Light Source (LCLS). The experiment employed a 9.5 keV photon energy and a 50 nm Bi_2Te_3 film on sapphire. The time resolution achieved in the experiment is approximately 50 fs, accounting for the laser and X-ray pulse durations, following the necessary correction for timing jitter between them. Figure 7.1(b) displays the normalized time-dependent diffracted intensity $(\Delta I/I)$ of the (556) peak, which is sensitive to both the $A_{1g}^{(1)}$ (1.8 THz) and $A_{1g}^{(2)}$ (3.9 THz) modes, as shown by the Fourier transform in Fig. 7.1(c).

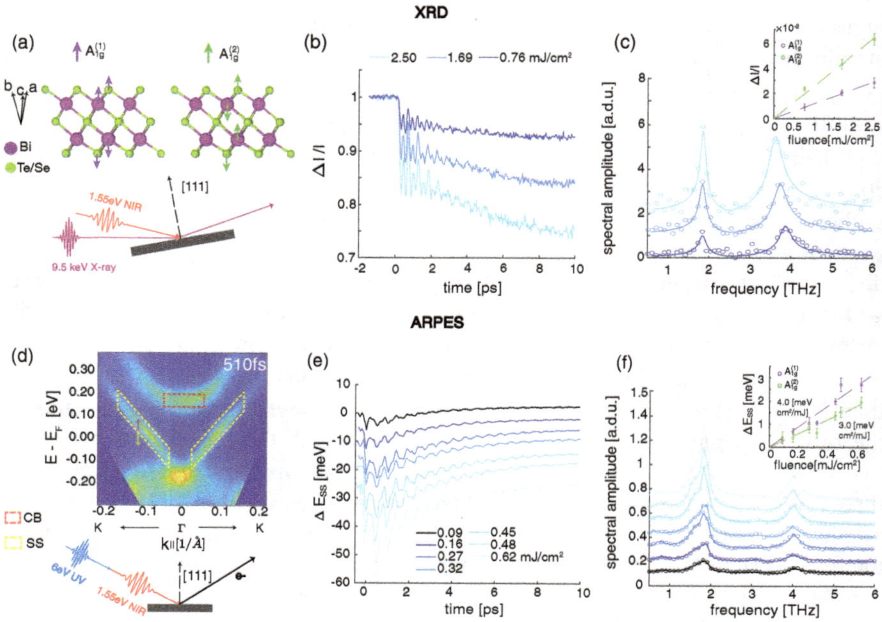

Fig. 7.1 Room temperature time-resolved measurements on Bi_2Te_3. (a–c) Time-resolved XRD. (a) shows two A_{1g} phonon modes of Bi_2X_3(X=Se,Te), in purple ($A_{1g}^{(1)}$) and green ($A_{1g}^{(2)}$) arrows, illustrating direction of atomic motion (along [111]) upon photoexcitation of the 1.55 eV laser. In the experiment, X-ray and NIR pulses are both grazing incident to match their penetration depth. Sample normal [111] is demonstrated with broken arrow. (b) Time-resolved diffraction of Bi_2Te_3 (556) peak under different fluences. Data taken at LCLS. (**c**) Fourier transform of (b). Inset shows linear fluence dependence of $A_{1g}^{(1)}$ and $A_{1g}^{(2)}$ mode induced (556) peak diffraction amplitudes $\Delta I/I$. (d–f) Time-resolved ARPES. (d) The electron population at the delay of 510 fs. The boxed area corresponds to a region of interest for the surface state (SS, yellow box) and conduction band (CB, red box). In the ARPES setup, the incident angles of the NIR pump and UV probe are both 50°. The A_{1g} phonon-induced surface state shift directions can be obtained from the initial phase of the coherent oscillation, as indicated by purple($A_{1g}^{(1)}$) and green($A_{1g}^{(2)}$) arrows. (e) Averaged energy shift of SS (ΔE_{SS}) in the yellow region of interest in (d). (f) Fourier transform of (b). Inset shows linear fluence dependence of SS energy modulation amplitude ΔE_{SS} induced by $A_{1g}^{(1)}$ and $A_{1g}^{(2)}$ mode. Black arrows point to the main peak identified as the bulk phonon coupled to SS

The excitation of A_{1g} phonons occurs through displacive excitation of coherent phonons (DECP), where the quasi-equilibrium phonon coordinates experience a sudden shift upon photoexcitation. This excitation induces phonon motion along the [111] direction, as illustrated by the purple ($A_{1g}^{(1)}$) and green ($A_{1g}^{(2)}$) arrows in Fig. 7.1(a). The time-dependent intensity modulations observed can be described as damped harmonic oscillators, each characterized by amplitude, frequency, damping constant, and initial phase. To extract these parameters, we employ a linear prediction singular value decomposition (LPSVD) method. The initial phases of modes are approximately either 0 or π, subject to the uncertainty in time-zero (which is approximately the quoted time-resolution for each experiment). Using the method similar to ref [1], we are able to annotate the atomic motion as shown in Fig. 7.1(a). The fact that the initial phases of atomic displacements are approximately either 0 or π implies that the photoexcited electron density produces forces, which are step-function-like in time, on the atoms. The fluence dependence of the $\Delta I/I$ amplitude is shown in the inset of Fig. 7.1(c), with the linear fits (represented by broken lines in the inset) constrained to intersect the origin. This constraint is based on the assumption that the coherent atomic motion amplitude is zero in the absence of optical excitation. By optimizing the phonon eigenvector over multiple Bragg peaks, the intensity modulation amplitude can be converted to the vibrational mode amplitude (see Appendix).

To determine the electron band energy shifts resulting from coherent phonon motion, we conducted time-resolved ARPES measurements on Bi_2Te_3 in a laser laboratory setting. A single crystal of Bi_2Te_3 was cleaved under ultrahigh vacuum conditions and probed using a 6 eV UV laser, achieving an overall time resolution of 60 fs. This quoted resolution is the full-width half maximum of the Gaussian, which represents the cross-correlation between the NIR pump and UV probe pulses.

Figure 7.1(d) presents the photoexcited ARPES spectrum at a delay time of 510 fs. The surface state (SS) is depicted in yellow, while the conduction band (CB) is represented in red, both indicated by boxed regions. To extract the energy distribution curve (EDC) peak position as a function of delay and parallel momentum $\mathbf{k}_{||}$ for the SS along Γ-K, we performed fitting. Negligible dependence on $\mathbf{k}_{||}$ was observed, allowing us to average the time-resolved energy shifts over $\mathbf{k}_{||}$ within the indicated region. This averaging maximizes the signal-to-noise ratio for subsequent analysis. It is important to note that the bulk states disperse in three dimensions, leading to significant uncertainty in the probed out-of-plane momentum component $\mathbf{k}_{\perp}$, which hinders a quantitative analysis [35]. In contrast, the SS lacks a $\mathbf{k}_{\perp}$-component and is therefore not affected by this ambiguity. Figure 7.1(e) displays the $\mathbf{k}_{||}$-averaged energy shifts of the SS for various fluences. The Fourier transform in Fig. 7.1(f) reveals two primary peaks, $A_{1g}^{(1)}$ and $A_{1g}^{(2)}$ (indicated by black arrows), along with shoulders associated with the surface termination [36, 37]. The main peaks exhibit the same frequency as the modes coupling to the bulk conduction band [37], suggesting that they correspond to bulk-like $\mathbf{q} = 0$ modes, allowing for a comparison with the modes measured by XRD. The amplitudes obtained from LPSVD are found to be linearly dependent on the fluence, as illustrated in the inset of Fig. 7.1(f). To validate the results of LPSVD, we compared the fitted spectra with the real and imaginary parts of the measured Fourier spectrum, and a good agreement was observed. Note that as in time-resolved XRD, the phases of each coherent oscillatory component are approximately either 0 or π up to uncertainty of time zero, which suggests that electron band shifts follow adiabatically the ionic motion, consistent with Born-Oppenheimer approximation. The electron band shift directions can be annotated as in Fig. 7.1(d). The fact that both coherent atomic motion and coherent electron band shifts have an initial phase close to 0 or π allows us to determine the sign of DPs as defined in Eq. 7.1.

For measuring the deformation potential of Bi_2Se_3, diffraction measurements were conducted at SACLA-FEL for both Bi_2Se_3 and Bi_2Te_3. A 9.5 keV hard X-ray source with a time resolution of 50 fs was employed in this experiment. In the case of Bi_2Se_3, a 90 nm film was deposited on BaF_2, while for Bi_2Te_3, a 200 nm film was deposited on sapphire. The two thin film samples were mounted side by side, ensuring that their surface normals were aligned within an accuracy of approximately 0.01°. This precise alignment minimizes systematic errors in fluence calibration between the two materials, ensuring the reliability of the comparative analysis.

Figure 7.2(a) presents the raw data (dots) and fits (solid lines) of the diffraction intensities for the (445) Bragg peak of Bi_2Te_3 and Bi_2Se_3. The data is grouped

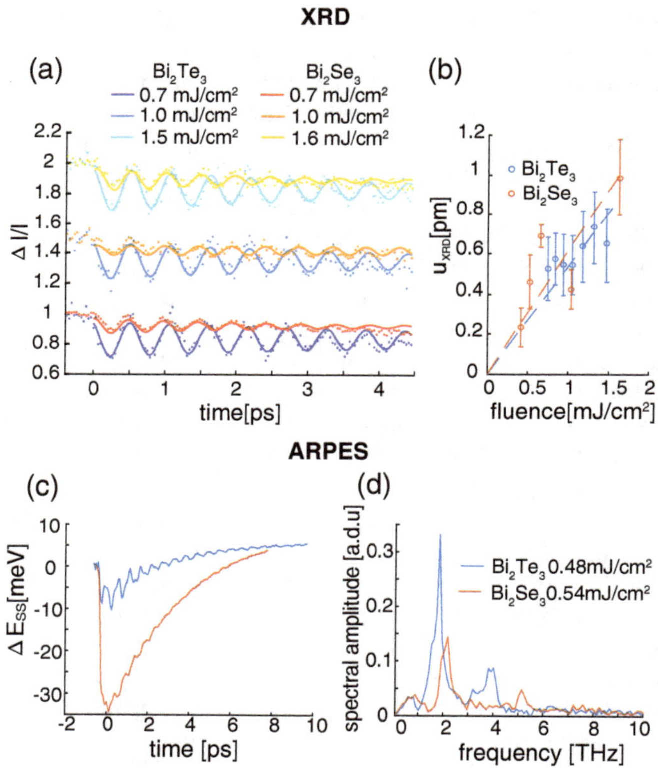

Fig. 7.2 Room temperature time-resolved measurements comparing photoexcited Bi_2Te_3 and Bi_2Se_3. (a) Bi_2Te_3 and Bi_2Se_3 time-resolved XRD on peak (445) measured at SACLA. The data are grouped into three levels of similar absorbed fluences. The intensity modulation of Bi_2Te_3 is consistently around three times larger than Bi_2Se_3. (b) Linear fluence dependence of u^{Bi}. The similar slopes of linear fits on Bi_2Te_3 and Bi_2Se_3 show that their Bi atoms are driven at similar amplitudes. (c) Bi_2Te_3 and Bi_2Se_3 time-resolved ARPES measurements of SS under similar absorbed fluences. (d) is the Fourier transform of (c)

based on similar absorbed fluence, taking into account differences in film thickness and optical absorption, and an offset is applied for clarity. Since the $A_{1g}^{(2)}$ mode was not resolved in Bi_2Se_3, only the atomic motion of the $A_{1g}^{(1)}$ mode is shown.

From the plot, it can be observed that the diffraction intensity modulations corresponding to the $A_{1g}^{(1)}$ mode in Bi_2Te_3 are consistently approximately three times larger than in Bi_2Se_3. Additionally, the sensitivity of the (445) peak to $A_{1g}^{(1)}$ distortion is about three times larger in Bi_2Te_3 compared to Bi_2Se_3 due to the differences in their eigenvectors. Consequently, the computed displacements u_{XRD} of the $A_{1g}^{(1)}$ mode in the two materials are concluded to be similarly driven, as shown in Fig. 7.2(b), where the slope of the linear fit (broken lines) for Bi_2Te_3 is approximately 0.88 times lower than for Bi_2Se_3.

The induced energy shifts of the surface states (SS) in Bi_2Se_3 were also measured using time-resolved ARPES experiments conducted under the same conditions as Bi_2Te_3, with a slightly different time resolution of 85 fs. Figure 7.2(c) displays the SS energy shifts for both materials under similar absorbed fluences. The Fourier transforms in Fig. 7.2(d) reveal that Bi_2Se_3 exhibits $A_{1g}^{(1)}$ (2.2 THz) and $A_{1g}^{(2)}$ (5.2 THz) phonons, similar to Bi_2Te_3. The frequencies of the A_{1g} phonons in Bi_2Se_3 are higher compared to Bi_2Te_3, primarily due to the lower average atomic mass.

Analysis of the ARPES data indicates that while the overall band shift is significantly larger in Bi_2Se_3, the oscillatory component corresponding to the $A_{1g}^{(1)}$ mode in Bi_2Te_3 is approximately 2.05 times larger than in Bi_2Se_3 when subjected to the same fluence.

Though we only focus on the coherent components of the time-resolved ARPES signal, we offer several possible reasons for the incoherent background (the exponential decay). A possible contribution to band-shifting dynamics is from surface photovoltage, which generates long-range electric fields that couple to the photoelectron trajectories following photoexcitation. The other contribution may be from electronic heating effects and the ensuing coupling of electrons to other excitations, including electron-phonon coupling channels. As the electrons cool they can generate phonons throughout the Brillouin zone in an incoherent fashion. These effects can be materials dependent.

7.3 Calculation of Deformation Potentials

The section will first discuss the calculations of Bi_2Te_3 DPs, based on a numerical simulation accounting for all intrinsic and extrinsic effects. We separately consider each effect below in order to provide the reader with intuition about the various ingredients and how to estimate their relative contributions. We will then discuss the calculations of Bi_2Se_3 DP, which is based on the known results of Bi_2Te_3.

Calculations for Bi_2Te_3*: Intrinsic Effects*

To compute the DP from experimental quantities, it is critical that the measurements are performed under the same incident fluence F. Pragmatically, it is convenient to measure the linear F-dependence of the displacements and energy shifts and use the fitted slopes to determine the DP:

$$\mathrm{DP} = \frac{(d\Delta E_{\mathrm{SS}}/dF)}{(du_{\mathrm{SS}}/dF)} \tag{7.2}$$

ARPES measures directly the SS energy shift (ΔE_{SS}), while XRD measures a weighted average of the atomic displacements over the probed region of the sample ($u_{\mathrm{XRD}} \neq u_{\mathrm{SS}}$) (recall that we defined u_{SS} in Eq. 7.1 as a weighted average over the surface state wave function extension). This brings errors intrinsically related to the nature of the measurements and must be addressed before Eq. 7.2 can be applied to compute the DP.

To address this, we must rescale the XRD measurements to give the atomic displacements in the vicinity of the SS:

$$u_{\mathrm{SS}} = \left(\frac{\tilde{u}_{\mathrm{SS}}}{\tilde{u}_{\mathrm{XRD}}}\right) u_{\mathrm{XRD}} \tag{7.3}$$

Quantities with tildes are obtained from simulations, while quantities without tildes are obtained directly from experimental measurements. The factor $\left(\frac{\tilde{u}_{\mathrm{SS}}}{\tilde{u}_{\mathrm{XRD}}}\right)$ is the simulated ratio between the average displacement $\tilde{u}_{\mathrm{SS}}$ over the spatial extent of the SS and the average displacement $\tilde{u}_{\mathrm{XRD}}$ measured by XRD. As we shall show, this ratio can be calculated by modeling the generation process of the depth-dependent coherent phonon field $u(z, t)$.

To ensure conceptual clarity, we will initially neglect the effects of diffusion and recombination in our analysis. We will revisit these factors later in this section.

The depth dependence of XRD measurements is determined by the X-ray penetration depth, which, under our experimental conditions, is 55 nm. In contrast, the wave function of the SS only extends approximately 1 nm from the surface [38]. These characteristic length scales can be represented as $1/\alpha_i$, where (i = XRD) represents the X-ray penetration depth at the specific incident angle in an XRD measurement and (i = SS) represents the spatial extent of the SS wave function. In XRD, the signal, i.e., the change in scattering intensity, is considered as a weighted average of the phonon amplitude over the region being probed. Similarly, the energy oscillation of the SS is determined by averaging the displacements over the extent of its wave function. These weighted averages can be written:

$$\tilde{u}_i(t) = \frac{\int_0^{L_i} dz \exp(-\alpha_i z) u(z, t)}{\int_0^{L_i} dz \exp(-\alpha_i z)} \tag{7.4}$$

where L_i represents the thickness of the sample. To compute the displacements $u(z, t)$, we model $u(z, t)$ as a harmonic oscillator displaced to a new quasi-equilibrium coordinate $u'(z, t)$ [39]:

$$\frac{d^2u}{dt^2} = -(2\pi f)^2(u(z,t) - u'(z,t)) - \gamma\dot{u}(z,t) \tag{7.5}$$

where f is the frequency and γ is the damping rate. The displaced coordinate $u'(z, t)$ is proportional to the photoinduced carrier density $n(z, t)$, which has the initial condition:

$$n(z, t = 0) \propto \exp(-\alpha_0 z) \tag{7.6}$$

where α_0 ($1/\alpha_0 = 15$ nm) is the absorption coefficient of the pump laser. Neglecting diffusion and recombination, the carrier density is static, and Eq. 7.6 is applicable for all times $t > 0$, that is, $n(z, t > 0) = n(z, t = 0)$. From Eqs. 7.4 and 7.6, we obtain:

$$\tilde{u}_i = u_0 \frac{\alpha_i}{\alpha_i + \alpha_0} \frac{1 - e^{-(\alpha_0+\alpha_i)L_i}}{1 - e^{-\alpha_i L_i}} \tag{7.7}$$

where u_0 represents the displacement at $z = 0$. We find that the rescaling factor is $\left(\frac{\tilde{u}_{SS}}{\tilde{u}_{XRD}}\right) = 2.6$, which shows that XRD probes a deeper region in which the weighted average displacements are substantially attenuated compared to the surface.

We now examine the possible role of carrier diffusion and recombination, which redistribute carriers following photoexcitation and can therefore introduce a time and depth dependence to the phonon driving forces. To motivate this discussion, we show a back-of-the-envelope estimate of the relevance of diffusion: using the carrier diffusivity of Bi_2Te_3 of $D_e = 520\,\text{nm}^2/\text{ps}$ [40], we can estimate a characteristic diffusion timescale of $1/(D_e\alpha_0^2) \approx 0.4$ ps. This is comparable with the phonon period for $A_{1g}^{(2)}$ of 0.25 ps. We conclude that diffusion redistributes carriers on a timescale comparable to that of the coherent phonon generation process, and its effect on the driving forces should be investigated.

To quantitatively model the effect, we retain Eq. 7.6 as an initial condition but allow the distribution to evolve via carrier diffusion along the sample normal direction $\hat{z}$:

$$\frac{dn}{dt} = D_e \frac{d^2n}{dz^2} - \frac{1}{\tau_0} n \tag{7.8}$$

where τ_0 is the carrier lifetime, which is estimated to be 20 ps [41]. We must also apply a boundary condition

$$\frac{dn}{dz}(z = 0) = 0, \frac{dn}{dz}(z = L_i) = 0 \tag{7.9}$$

to enforce no diffusion across the sample boundaries at 0 and L_i.

After solving the partial differential equation, Eq. 7.8, we obtain the carrier density profile $n(z,t)$. To connect this to our experiment, we can then compute the coherent phonon amplitude $u(z,t)$ from Eq. 7.5 by taking $u'(z,t) \propto n(z,t)$ as before, but now with the time dependence included. To accurately model the experiment, we must also consider that the phonon frequency is influenced by $n(z,t)$, as evidenced by the fluence-dependent mode softening shown in Fig. 7.9(a). We incorporate this effect by assuming a linear dependence of the mode frequency $f(z,t)$ on fluence: $f(z,t) = f_0(1 - \beta n(z,t))$. Here, f_0 represents the equilibrium frequency, and $\beta > 0$ is a proportionality constant that determines the magnitude of the softening. By making this assumption, we can numerically solve for $u(z,t)$ and subsequently calculate the average displacements $\tilde{u}_i(t)$ using Eq. 7.4.

In our simulation of Bi_2Te_3, we focus on the $A_{1g}^{(2)}$ mode and consider a value of $D_e = 520\,\mathrm{nm}^2/\mathrm{ps}$ [40]. By choosing an appropriate β parameter, we aim to reproduce the observed linear softening. The damping constant γ is set to 0.25 ps^{-1} to align with the experimental data. Running the simulation, we find that $\left(\frac{\tilde{u}_{SS}}{\tilde{u}_{XRD}}\right) = 2.2$ for the $A_{1g}^{(2)}$ mode, while the corresponding factor for the $A_{1g}^{(1)}$ mode is 2.0. Comparing these results to the factor of 2.6 computed earlier in Eq. 7.7 without considering diffusion and recombination, we observe that the carrier dynamics introduce a correction of approximately 15% on top of the considerations related to the penetration depth. Therefore, for precise quantitative analysis, it is crucial to account for diffusion and recombination. Nevertheless, it is worth noting that the primary correction arises from the substantial difference in the penetration depth of XRD measurements compared to the extension of the surface state wave function.

Calculations for Bi_2Te_3*: Extrinsic Effects*

We now return to the issue of extrinsic errors, one of which is associated with fluence miscalibration across different experimental facilities. The other error is due to different time resolutions τ in ARPES and XRD measurements.

The fluence calibration can vary substantially between different experimental labs: in published Bi measurements, the reported fluences for the same amount of phonon softening differ by up to a factor of 5 in independent measurements [42–45]. These discrepancies may be attributed to miscalibration in the laser power measurement, or errors in the measurement of the incident area on the sample. Therefore, for quantitative calculation of the DP, we must investigate the possibility of fluence miscalibration between our XRD and ARPES measurements.

Errors in fluence can be calibrated by using $A_{1g}^{(2)}$ frequency as a proxy, due to its strong dependence on fluence. Explicitly, we need to define

$$u_{SS} = \frac{(d\tilde{u}_{SS}/d\tilde{f}_{SS})}{(d\tilde{u}_{XRD}/d\tilde{f}_{XRD})} u_{XRD} \tag{7.10}$$

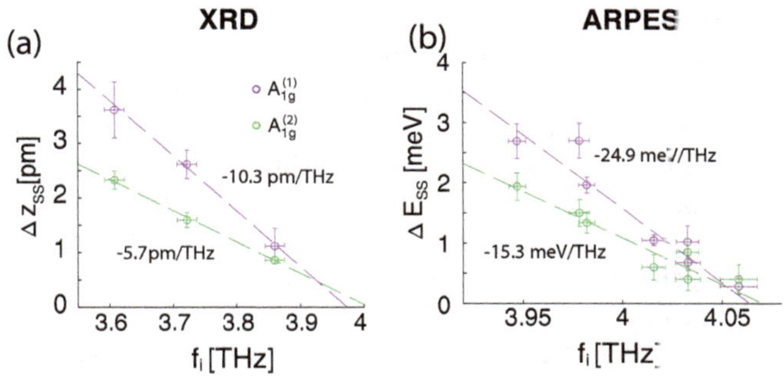

Fig. 7.3 (a) The Bi_2Te_3 u_{SS} as a function of the measured $A_{1g}^{(2)}$ frequency. (b) SS band shift ΔE_{SS} as a function of the measured $A_{1g}^{(2)}$ frequency

in place of Eq. 7.3, and accordingly

$$\mathrm{DP} = \frac{(d\Delta E_{SS}/df_{SS})}{(du_{SS}/df_{XRD})} \tag{7.11}$$

where f_i is the $A_{1g}^{(2)}$ frequency measured by the respective techniques. We note that $f_{SS} \neq f_{XRD}$ even for the same incident fluence F. This is again due to the depth dependence of phonon fields and the different sensitivities of ARPES and XRD as described by Eq. 7.4. Equation 7.11 is mathematically equivalent to the original definition (Eq. 7.2). By doing so, errors in fluence calibration are effectively circumvented by using the softened frequency as a proxy.

Note that both of the measured quantities u_{XRD} and ΔE_{SS} should be multiplied by an additional minor correction due to limited time resolution τ (60 fs in ARPES and 50 fs in XRD measurements), estimated as below by convolving a sine wave at the phonon (both A_{1g} modes) frequency f_0 with a Gaussian with full-width-half-max τ:

$$\exp\left(\frac{\pi^2 f_0^2 \tau^2}{4\ln 2}\right) \tag{7.12}$$

Equation 7.11 can then be used to calculate DPs of A_{1g} modes in Bi_2Te_3. In Fig. 7.3(a), we plot u_{SS} as a function of $A_{1g}^{(2)}$ frequency and ΔE_{SS} as a function of $A_{1g}^{(2)}$ frequency. The slopes of these linear fits are used to calculate DPs in Eq. 7.11. The resulting DPs are listed in Table 7.2, which reflects our best estimate considering all intrinsic and extrinsic factors described to this point. We note that these values differ by 20% compared with the DP calculation results without using frequency as a proxy for fluence, suggesting a miscalibration of 20% between the two measurement facilities.

The error bars in the table are determined through the following procedures. Statistical errors are derived from the fitting error bars in Fig. 7.3(a), (b), taking into account both the horizontal and vertical error bars.

Systematic errors are estimated by considering the uncertainties associated with two key parameters in our calculations: the X-ray incidence angle (and consequently the penetration depth) and the magnitude of the diffusivity. Regarding the incidence angle, we estimate an uncertainty of $\pm 0.1°$ with respect to the nominal angle of $0.5°$, resulting in penetration depths ranging from 35 to 70 nm. This introduces an error of 8% in the calculation of the DP.

To estimate the systematic error associated with the diffusivity, we varied its value from 200 to 1000 nm^2/ps, which deviates by approximately a factor of 2 from the estimated value of $D_e = 520\,\text{nm}^2/\text{ps}$. This introduces a corresponding uncertainty of 15% in the DP determination. We note that the above assumes only majority carrier diffusivity, but the ambipolar diffusivity falls well within the range of our uncertainty [46]. Although there can be a change in the diffusivity upon photoexcitation [47, 48], we expect it to quickly return to its near equilibrium level as the photo-excited carriers relax.

Calculation for Bi_2Se_3

To compute the DP of Bi_2Se_3, we cannot directly employ the strategies from Section II.B that minimize the systematic errors due to fluence calibration. The $A_{1g}^{(2)}$ mode of Bi_2Se_3 was not observed by XRD and is therefore not available to use as a proxy for fluence. Instead, we exploit the fact that Bi_2Se_3 and Bi_2Te_3 were measured side by side, thus allowing us to compute the ratio of their DPs as defined in Eqs. 7.3 and 7.2:

$$\text{DP} = \left(\frac{\tilde{u}_{\text{SS}}}{\tilde{u}_{\text{XRD}}}\right)^{-1} \frac{(d\Delta E_{\text{SS}}/dF)}{(du_{\text{XRD}}/dF)} \tag{7.13}$$

Based on the XRD measurements at SACLA and the ARPES measurements comparing Bi_2Te_3 and Bi_2Se_3 (as depicted in Fig. 7.2), we determine that the ratio $\frac{(d\Delta E_{\text{SS}}/dF)}{(du_{\text{XRD}}/dF)}$ between Bi_2Te_3 and Bi_2Se_3 is 2.33. However, we must consider the influence of different carrier diffusivity values in the two materials, with Bi_2Se_3 having a much higher value ($D_e = 1400\,\text{nm}^2/\text{ps}$) [49]. As a result, diffusion leads to a lower carrier density near the SS in Bi_2Se_3 compared to Bi_2Te_3, resulting in a reduced $\tilde{u}_{\text{SS}}$ and an additional correction factor of 0.63 to the DP ratio between Bi_2Te_3 and Bi_2Se_3 due to $\left(\frac{\tilde{u}_{\text{SS}}}{\tilde{u}_{\text{XRD}}}\right)$.

Taking into account the correction factor due to time resolution differences, which has already been considered and yields a factor of 0.93, the final DP ratio of Bi_2Te_3 over Bi_2Se_3 is 1.37.

Table 7.1 Deformation potential [meV/pm]. The experiment deformation potentials of the surface states are average over the region of interest across $\mathbf{k}_{||}$ space as shown in Fig. 7.1(d). The experiment values take into account both intrinsic and extrinsic corrections. The theory values are DPs of Γ point SS. Error bars include both systematic and statistical errors

Variable	Experiment	Theory (LDA)
Bi_2Te_3, $A_{1g}^{(1)}$	2.42 (0.74)	5.8
Bi_2Te_3,$A_{1g}^{(2)}$	2.67 (0.77)	6.8
Bi_2Se_3, $A_{1g}^{(1)}$	1.77 (0.64)	1.2

With the knowledge of the $A_{1g}^{(1)}$ DP ratio, we can then calculate the Bi_2Se_3 DP based on the known Bi_2Te_3 DP, as presented in Table 7.1. The uncertainties include both statistical error (32%) and systematic error (16%) arising from uncertainties in X-ray incidence angle calibration and carrier diffusivities.

The non-observation of Bi_2Se_3 $A_{1g}^{(2)}$ mode in the SACLA XRD measurement is attributed to the limited time resolution and the considerable X-ray beam intensity fluctuations. Based on the peak sensitivity and the noise level, we estimate an upper bound of the $A_{1g}^{(2)}$ mode atomic displacement in Bi_2Se_3, and thus lower bound of the Bi_2Se_3 $A_{1g}^{(2)}$ DP of 0.2 meV/pm.

7.4 Benchmarking the Experimental Method and Theoretical Calculations

We performed DFT calculations to compare with the experimentally measured A_{1g}-SS DPs of Bi_2Te_3 and Bi_2Se_3. See Appendix for details. The computed electronic band structures of Bi_2Te_3 and Bi_2Se_3 are shown in Fig. 7.4(a) and (b). In our electronic structure calculations, we use the vacuum level of the Hartree potential to align the energies of the electronic states.

DPs are obtained using the frozen phonon method, where the atoms are displaced along the A_{1g} mode coordinates (e.g., as defined in Fig. 7.1(c)), and the SS energy shifts are calculated based on the resulting shifts in the electronic band structure of a slab (e.g., as defined in Fig. 7.1(d)). Our frozen phonon calculation assumes the Born-Oppenheimer approximation which is fully compatible with the experimental observation. We assume that bulk phonon modes are homogeneously distributed along the slab despite the presence of the surface. The computed DPs along K-Γ-M are shown in Fig. 7.4(c) and (d).

As observed in the experiment, the $A_{1g}^{(1)}$ DPs are weakly $\mathbf{k}_{||}$-dependent near the zone center (see Appendix). However, the strong $\mathbf{k}_{||}$-dependence of the $A_{1g}^{(2)}$ DPs in theory does not agree well with experiment.

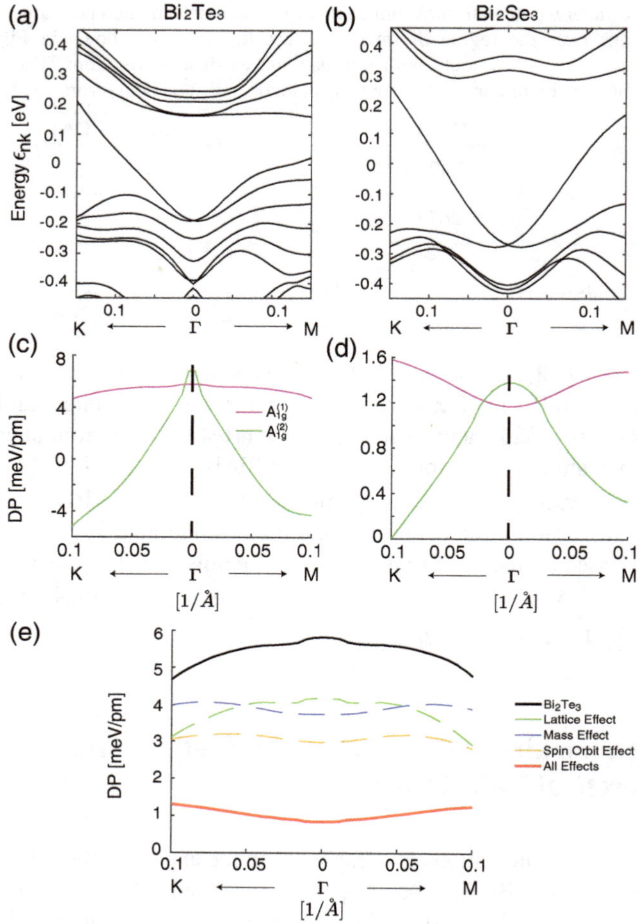

Fig. 7.4 Density functional theory calculation of electronic band structures and deformation potentials. (a) Band structure for a 6-quintuple layer (QL) slab of Bi_2Te_3, shown for part of the reciprocal space. (b) Band structure for a 5-QL slab of Bi_2Se_3. (c) Deformation potential (DP) of the two A_{1g} modes in Bi_2Te_3 derived from frozen phonon calculations. (d) Deformation potentials of the A_{1g} modes in Bi_2Se_3. (e) Calculated A_{1g} mode DP by swapping the lattice constants' value, masses of group VI atoms, and spin-orbit coupling strength in Bi_2Te_3 with that of Bi_2Se_3. Combining all three effects, the six times larger A_{1g} mode DP in Bi_2Te_3 can be reasonably explained

We average the computed DP values near the Γ-point to compare with the experimental values and include them in Table 7.1. The calculated $A_{1g}^{(1)}$ DP in Bi_2Se_3 is 1.5 times smaller than in the experiment and agrees within the experimental uncertainties. In Bi_2Te_3, the A_{1g} DPs are about 2.5 times larger than the experimental values and outside of the confidence interval of measurements. We emphasize that this comparison does not reveal an order of magnitude difference and consider the theoretical and experimental difficulties to obtain precise and accurate values on a scale of a few meV per unit cell. In this context, a difference of factor 2 should be understood as general agreement. We also note the consistent signs of experimentally derived and theoretically obtained DPs.

From the perspective of theory, it is noteworthy that our calculations predict a 4.8× difference between the $A_{1g}^{(1)}$ DPs for the two materials, which was not observed in an experiment where values differ by only 1.4. This discrepancy might be attributed to limitations of DFT calculations in capturing the precise band structure, which might also lead to the inconsistency regarding the $\mathbf{k}_{||}$-dependence of the DPs. It has been shown that GW calculations, which account for electron correlations, show a better agreement with the band structure measured by ARPES [50].

To further investigate possible mechanisms underlying the computational discrepancies, we conducted a numerical experiment in which we modify individual parameters that are different in the two systems. These are lattice constants, Se and Te atomic masses, and spin-orbit strength. We start with Bi_2Te_3 and evaluate the contribution of each individual parameter of Bi_2Se_3 to the DPs of Bi_2Te_3. Note that we only varied one parameter at a time while holding the rest at their nominal values for Bi_2Te_3. The resulting $A_{1g}^{(1)}$ deformation potentials are plotted in Fig. 7.4(e). If the Se atomic mass replaces the Te atomic mass, the DP is reduced by ~35% due to the change in eigenvector (see Table 7.2). By modifying the spin-orbit interaction strength of Te to the value in Se, the DP is reduced by ~50%. When the Bi_2Se_3 lattice parameters are used for the Bi_2Te_3 system, the DP at Γ is reduced by ~27%, which is mostly related to the lattice expansion effects on the crystal field splitting. Finally, when all these factors are simultaneously incorporated, Bi_2Te_3 is effectively transformed into Bi_2Se_3, resulting in the 4.8× reduction mentioned previously. These calculations demonstrate that no single factor is sufficient to explain the difference in the calculated DPs of Bi_2Te_3 and Bi_2Se_3 and that lattice constants, atomic masses, and spin-orbit coupling strength are all key ingredients.

Additionally, we performed a constrained DFT calculation (see Appendix) of the vibrational modes in the photoexcited states. We found that for Bi_2Te_3, the mode amplitude dependence on $A_{1g}^{(1)}$ frequency reproduces the measurement well and that the ratio between Bi_2Te_3 and Bi_2Se_3 phonon amplitudes given the same excitation density is consistent with the experiment. In general, DFT captures well how photoexcited electrons couple to the phonon modes in these materials.

To summarize, using time-resolved XRD and ARPES measurements on Bi_2Te_3 and Bi_2Se_3, we extract the DPs that quantify the coupling between topological surface states and A_{1g} phonon modes. We perform a comprehensive analysis of correction factors due to intrinsic and extrinsic errors, when integrating the results of XRD and ARPES measurements. For intrinsic effect corrections, the most substantial contribution is due to the different depth sensitivities of the two probes. More modest corrections are necessary to account for photoexcited carrier diffusion and recombination. For extrinsic corrections, we showed that fluence miscalibration can be circumvented by using the strongly fluence-dependent $A_{1g}^{(2)}$ frequency in

Bi_2Te_3. DFT calculations show that the larger $A_g^{(1)}$ DP in Bi_2Te_3 compared to Bi_2Se_3 can be attributed to their differences in lattice constants, atomic masses, and spin-orbit coupling strength which makes the largest contribution.

We believe that the results which quantify the electron-SS interaction can be useful for understanding SS transport, and the methodology we established can be insightful for studying the structure-induced topological switch.

We point out, though, that the limitations of our method can be lifted in future experiments. First, extensive modeling was required to arrive at values for the DP from the experimentally measured quantities. This introduces substantial complexity to the analysis and couples in uncertainties due to the dependence on material parameters such as penetration depth and diffusivity. We suggest that future experiments be performed on thin films (thinner than the penetration depth) to ensure homogeneous carrier distribution after excitation regardless of the probe penetration depth. This approach was taken in ref. [33], resulting in only a modest factor of 1.03 to account for the different penetration depths of ARPES and XRD. We also point out that our analysis is focused on the SS, the length scale of which dictates the depth sensitivity of the ARPES measurement. This would be different for bulk electronic states, in which the inelastic mean free path of photoelectrons determines the probing depth. Lower-temperature experiments are favorable as they reduce phonon damping and reveal more phonon components coupled to SS, which can enhance the accuracy of mode separation in our method. Finally, we note that experiments simultaneously combining ARPES and diffraction can in principle be realized on free-electron laser facilities, which could open a new experimental paradigm for studying electron-phonon coupling.

Appendix

Phonon Eigenvector Reconstruction

Figure 7.5 shows the crystal structure of space group $R\bar{3}m$ and provides a 1D-atom-chain view of $A_{1g}^{(1)}$ mode on the right side. The quintuple-layer view of $A_{1g}^{(1)}$ mode (same as Fig. 7.1(a)) is provided in the pink box. Different layer patterns are represented with A, B, and C. Atoms that form the 1D-atom-chain are all taken from the A layer. A primitive unit cell is outlined with gray solid lines. Among the three types of atoms for A_{1g} modes, Bi, X(I), and X(II), the X(I) atoms are not allowed to displace in a A_{1g} mode, as these atoms are the centers of inversion symmetry. Thus, an A_{1g} phonon has only two degrees of freedom determined by Bi and X(II) atomic motion along [111]. Since there are two free parameters, there are correspondingly two A_{1g} modes.

We first refine the internal atomic coordinates using the static diffraction intensity on different peaks. Second, to determine the eigenvector, we use the diffraction intensity modulation induced by the same phonon mode measured across multiple

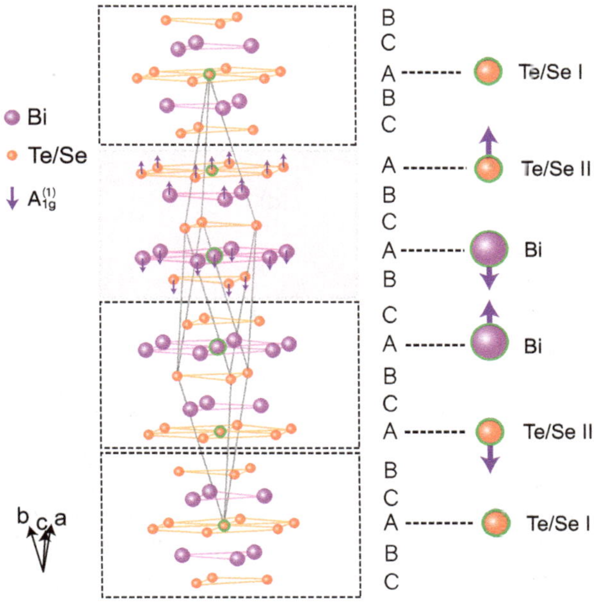

Fig. 7.5 Crystal structure of Bi_2X_3 (X = Se,Te) and schematic for $A_{1g}^{(1)}$ phonon mode

Bragg peaks. Finally, based on the measured eigenvectors, Δz_λ^{Bi} can be quantified from diffraction. We detail the second step below.

If we represent the X(II) atom coordinates in the primitive unit cell as $(\delta_1, \delta_1, \delta_1)$ and the Bi as $(\delta_2, \delta_2, \delta_2)$, the scattering structure factor,

$$F = f_X + 2e^{i\pi(h+k+l)}[f_X \cos[\pi(1-2\delta_1)(h+k+l)] + f_{Bi}\cos[\pi(1-2\delta_2)(h+k+l)]] \tag{7.14}$$

$\delta_1 \approx \frac{1}{5}$ and $\delta_2 \approx \frac{2}{5}$. For Bragg peaks, h, k, l are integers, and we let $h+k+l = N$. $N = 5n$ Bragg peaks will not be sensitive to either of the A_{1g} modes. $N = 5n \pm 1$ peaks have the same sensitivity to a given A_{1g} mode for a given n and are sensitive to both $A_{1g}^{(1)}$ and $A_{1g}^{(2)}$. $N = 5n \pm 2$ have the same sensitivity to a given A_{1g} mode for a given n and are sensitive mainly to only $A_{1g}^{(1)}$ mode.

During XRD experiment at LCLS, as we orient the sample by rotating it around the azimuthal axis (aligned to the sample surface normal to within 0.1°), we obtained the static X-ray scattering intensity on a total of seven peaks. This allows us to refine the lattice constants and the internal coordinates (δ_1, δ_2), which show good agreement with DFT calculations.

The eigenvector of Bi_2Te_3 A_{1g} phonon mode can be obtained from the intensity modulation amplitude of a set of peaks sensitive to the mode. We need only one

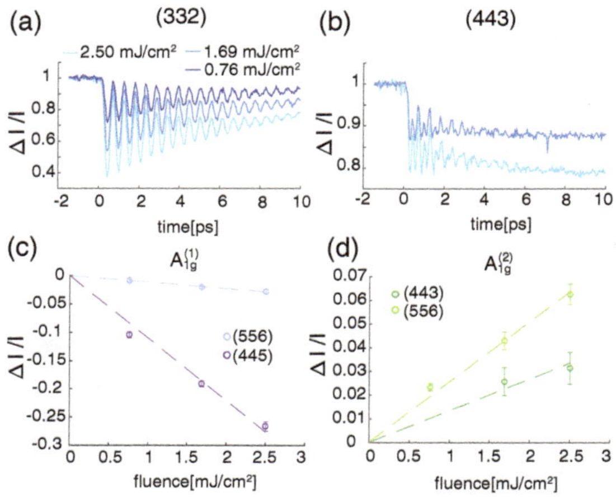

Fig. 7.6 LCLS XRD data for the eigenvector reconstruction. (a) and (b) The time-resolved diffraction on peaks (332) and (443). Peak intensity modulation amplitude induced by a given A_{1g} mode is different on peaks (332) and (443), which contains information for calculating A_{1g} eigenvector. (c) shows the $A_{1g}^{(1)}$ induced normalized intensity modulation amplitude $\Delta I/I$ on peaks (445) and (556). (d) shows the $A_{1g}^{(2)}$ induced intensity modulation amplitude on peaks (443) and (556)

free parameter to describe two A_{1g} mode eigenvectors simultaneously, and we pick this free parameter to be Bi to Te ratio in $A_{1g}^{(1)}$ mode. The experimental observables are the Bragg peak intensity modulations induced by certain A_{1g} mode (Fig. 7.6(a) and (b)). As shown in Fig. 7.6(c) and (d), amplitudes of the normalized Bragg peak intensity modulation ($\Delta I/I$) linearly increase with fluence. The increase in $\Delta I/I$ per unit fluence (1 mJ/cm^2) is the slope of the linear fit constrained through the origin in Fig. 7.6(c) and (d). We aim to solve for three unknown variables: (1) Bi to Te ratio in $A_{1g}^{(1)}$ mode, (2) $A_{1g}^{(1)}$ mode amplitude (alternatively one can also choose to use Δz^{Bi}) increase per the unit fluence, (3) $A_{1g}^{(2)}$ mode amplitude increase per unit fluence. We, therefore, look for three equations, and they are the linear slopes of (1) $A_{1g}^{(1)}$ modulated diffraction intensity of (556) versus fluence, (2) $A_{1g}^{(1)}$ modulated diffraction intensity of (445) versus fluence, and (3) $A_{1g}^{(2)}$ modulated diffraction intensity of (556) versus fluence. Note that to avoid near-degeneracy of the equation set, for $A_{1g}^{(1)}$ mode, we pick two peaks that satisfy either $N = 5n \pm 1$ or $N = 5n \pm 2$. The solved results have an agreement to DFT calculated eigenvector to within 15%, in terms of the *normalized* $A_{1g}^{(1)}$ eigenvector projected onto a Bi atom.

In Table 7.2, we show the normalized eigenvectors of Bi_2Te_3 calculated through XRD measurements and of Bi_2Se_3 obtained through DFT. For Bi_2Se_3, we did not resolve signals on a sufficient number of Bragg peaks to directly solve the eigenvectors.

Table 7.2 The normalized eigenvectors of Bi_2Te_3 and Bi_2Se_3 A_{1g} mode

Mode	Bi	Te/Se II
Bi_2Te_3, $A_{1g}^{(1)}$	0.44	–0.56
Bi_2Te_3,$A_{1g}^{(2)}$	0.56	0.44
Bi_2Se_3, $A_{1g}^{(1)}$	0.58	–0.40
Bi_2Se_3, $A_{1g}^{(2)}$	0.40	0.58

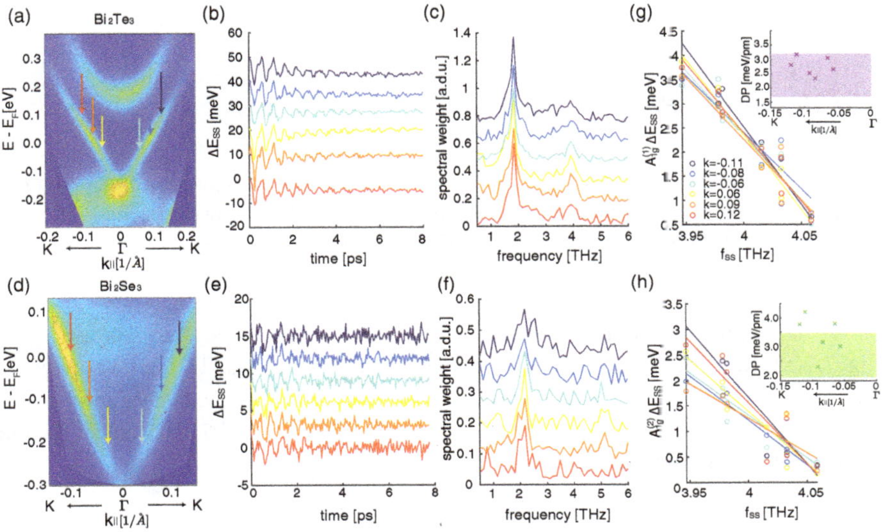

Fig. 7.7 **k**-dependence of time-resolved ARPES measurements. (a) The electron population of Bi_2Te_3 measured at 510 fs delay under 1.02 mJ/cm^2. The arrows show different $\mathbf{k}_{||}$ regions we sample for the surface state. (b) Surface state modulation of Bi_2Te_3 surface state under 1.02 mJ/cm^2, sampled at different $\mathbf{k}_{||}$ regions. (c) The Fourier transform of (b). (d) The electron population of Bi_2Se_3 measured at 510 fs delay under 0.54 mJ/cm^2. The arrows show different $\mathbf{k}_{||}$ regions we sample for the surface state. (e) Surface state modulation of Bi_2Se_3 surface state under 0.54 mJ/cm^2, sampled at different $\mathbf{k}_{||}$ regions. (f) The Fourier transform of (e). (g) Bi_2Te_3 $A_{1g}^{(1)}$ phonon-induced energy modulation amplitude as a function of $A_{1g}^{(2)}$ frequency. The slopes of the linear fits across $\mathbf{k}_{||}$ points do not vary much, which justifies averaging over a large region of interest in $\mathbf{k}_{||}$ space to improve the signal-to-noise ratio. The inset contains part of Fig. 7.4(c), with additional crosses representing the $A_{1g}^{(1)}$ DPs calculated for individual $\mathbf{k}_{||}$ points. (h) Bi_2Te_3 $A_{1g}^{(2)}$ phonon-induced energy modulation amplitude as a function of $A_{1g}^{(2)}$ frequency. Crosses in the inset represent the $A_{1g}^{(2)}$ DPs calculated for individual $\mathbf{k}_{||}$ points

k-*Dependence of Surface State Energy Shifts*

To enhance the signal-to-noise ratio of the surface electron band coherent oscillations, we average the change in band energy over ± 0.2 Å^{-1} near Γ, which is based on the fact that the DPs and thus the electron band shifts are nearly constant across this area of reciprocal space. In Fig. 7.7, we show the **k**-dependent surface state response; the sampled $\mathbf{k}_{||}$ points are illustrated in Fig. 7.7(a) and (d). In Fig. 7.7(g) and (h), the energy shift versus $A_{1g}^{(2)}$ frequencies for different $\mathbf{k}_{||}$ points are plotted for $A_{1g}^{(1)}$ and $A_{1g}^{(2)}$ in Bi_2Te_3. The $\mathbf{k}_{||}$ dependence of $A_{1g}^{(1)}$-induced $\Delta E^{n\mathbf{k}}$ is weak (variation of 13%). The $A_{1g}^{(2)}$ has a low signal-to-noise ratio and affects the amplitude extraction, leading to higher variations in DPs. Overall, averaging over a larger portion of **k**-space to improve the signal-to-noise ratio is justified.

Simulations of Phonon Fields and Probed Signals

As in the main text, we present the simulation details for DP calculations in Bi_2Te_3 first, before we do so for Bi_2Se_3. We show the simulated carrier density $n(z, t)$ and the phonon field $u(z, t)$ in Fig. 7.8, for both the thin film used in XRD and the bulk sample used in ARPES. This simulation is used to obtain the scaling factors in Eqs.7.3 and 7.10.

We show the measured fluence dependence of the mode frequencies f_i in Fig. 7.9(a), for different samples and experimental conditions. The large change of the Bi_2Te_3 $A_{1g}^{(2)}$ frequency with increasing fluence shows that f_i can be a good proxy for the absorbed pump energy density. Thus, we can use Eq. 7.11 to correct for the fluence uncertainty in different experiments.

The DPs are proportional to $[\frac{(d\tilde{u}_{SS}/d\tilde{f}_{SS})}{(d\tilde{u}_{XRD}/d\tilde{f}_{XRD})}]^{-1}$, and hence:

$$\left[\frac{(d\tilde{u}_{SS}/d\tilde{f}_{SS})}{(d\tilde{u}_{XRD}/d\tilde{f}_{XRD})}\right]^{-1} = \left(\frac{\tilde{u}_{SS}}{\tilde{u}_{XRD}}\right)^{-1} \times \frac{(d\tilde{f}_{SS}/dF)}{(d\tilde{f}_{XRD}/dF)} \tag{7.15}$$

The first factor in Eq. 7.15 right-hand side is presented in the main text, assuming a fixed fluence. Simulated fluence (F) dependence of $\tilde{f}_i$ in XRD (using thin film) and ARPES (using bulk) is shown in Fig. 7.9(b). $\tilde{f}_i$ is obtained from $\tilde{u}_i(t)$ as shown in Fig. 7.8 by LPSVD. The slopes of linear fits in Fig. 7.9(b) reproduce the experimental observation that the f_i softening for XRD is larger than ARPES. We use this to calculate the quantity $\frac{(d\tilde{f}_{SS}/dF)}{(d\tilde{f}_{XRD}/dF)}$ the second term of Eq. 7.15 right-hand side. It might seem counterintuitive that for a given fluence, the phonon softening is larger in an XRD measurement, while the phonon amplitude is larger in an ARPES measurement. This is because phonon amplitudes are quantified by their instantaneous behavior near $t = 0$. Diffusion effects are limited to the timescale of phonon generation of approximately one phonon period. In contrast, the analysis of phonon frequencies requires multiple phonon cycles. This allows for diffusion to act longer, which leads to the larger carrier density in a thin film sample as compared to the surface area of a bulk.

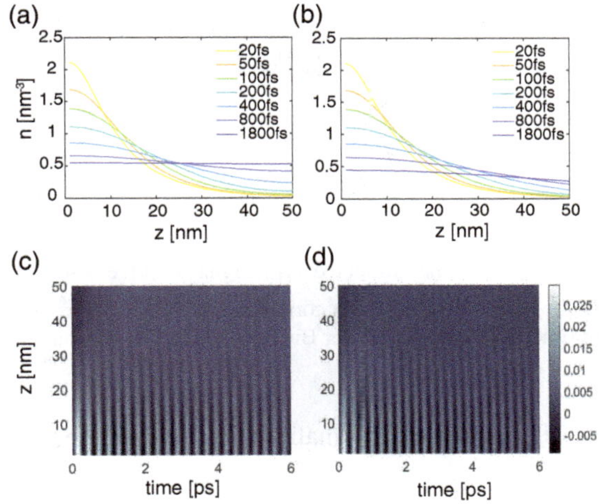

Fig. 7.8 Diffusivity simulation. Nonequilibrium carrier density profile for different time delays after the laser pump of fluence 2 mJ/cm^2, in a 50 nm thin film (a) and bulk (b). (c, d) Coherent phonon field induced by the population of nonequilibrium carriers, in a 50 nm thin film (c) which corresponds to (a), and bulk (d) which corresponds to (b)

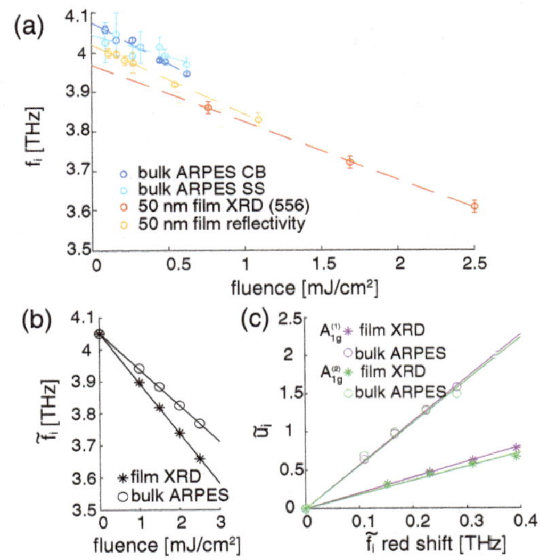

Fig. 7.9 (a) Linear fluence dependence of Bi_2Te_3 $A_{1g}^{(2)}$ mode frequency. We included data taken with XRD (50 nm film), ARPES (bulk), and pump-probe optical reflectivity (50 nm film), which span the different range of absorbed fluences and thus different range of f_i. (b) Simulated $A_{1g}^{(2)}$ mode frequency dependence on fluence. (c) The simulated A_{1g} normalized signals of time-resolved ARPES for bulk sample, and the normalized signals of time-resolved XRD for a 50 nm thin film sample (corresponds to LCLS experiment), as a function of $A_{1g}^{(2)}$ frequency redshift

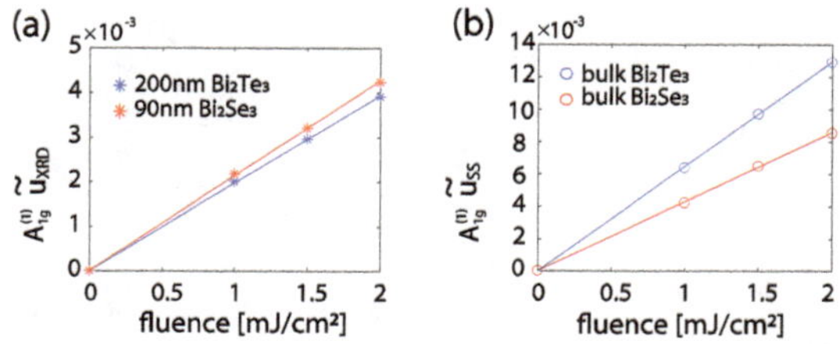

Fig. 7.10 (a) The simulated $A_{1g}^{(1)}$ signal of time-resolved XRD for Bi_2Te_3 film (200 nm) and Bi_2Se_3 film (90 nm), which correspond to the data taken at SACLA. (b) The simulated $A_{1g}^{(1)}$ signal of time-resolved ARPES for Bi_2Te_3 bulk and Bi_2Se_3 bulk

The simulated, normalized A_{1g} amplitudes as a function of $A_{1g}^{(2)}$ redshift are shown in Fig. 7.9(c). The slopes are used to calculate the quantity $\frac{(d\tilde{u}_{SS}/d\tilde{f}_{SS})}{(d\tilde{u}_{XRD}/d\tilde{f}_{XRD})}$. The slope of the ARPES measurement $(d\tilde{u}_{SS}/d\tilde{f}_{SS})$ is a factor of 2.8 (3.0) larger than the XRD measurement $(d\tilde{u}_{XRD}/d\tilde{f}_{XRD})$ for the $A_{1g}^{(1)}$ $(A_{1g}^{(2)})$ mode.

Next, we consider the carrier diffusion simulation for the DP calculation of Bi_2Se_3. Here, different samples are measured using the same experimental conditions and we can ignore the systematic errors induced by fluence calibration. In these simulations, the absolute values of $\tilde{u}$ are irrelevant, and only the ratios between them matter. Figure 7.10 shows that given the same absorbed fluence, the response of Bi_2Te_3 and Bi_2Se_3 differs, both for XRD in Fig. 7.10(a) and for ARPES in Fig. 7.10(b). Again, this is due to the difference in carrier diffusivities of Bi_2Te_3 and Bi_2Se_3.

Ellipsometry

We measured ellipsometry data to obtain the refractive index (real and imaginary parts) and in turn calculate the absorbed fluence in both Bi_2Te_3 and Bi_2Se_3 samples. The ellipsometer measures reflectivity with the phase of p and s polarized light. The ratio of them, a complex number, is expressed as $\frac{r_p}{r_s} = \tan \Psi e^{i\Delta}$. In Fig. 7.11, we plot Ψ and Δ as measured (dots) and as fitted (solid curve). The ellipsometry measurement was conducted under four different incidence angles. The fit uses a multilayer reflection model and takes into account the thickness of the film as well as the refractive index of the substrate. The complex refractive indices as fit results are shown in Fig. 7.11(e) and (f). To compare the pump energy density in different measurements using different incidence angles (XRD: 89.5°, ARPES:50°, optical: <5°, grazing angle is defined as the angle between the beam and the sample normal), we calculate the reflectivity as a function of incidence angle (Fig. 7.11(g)), under the center wavelength of the pump laser, 780 nm. In Fig. 7.11(h), we show the reflectivity of a 250 nm thick Bi_2Te_3 measured under normal incidence (with

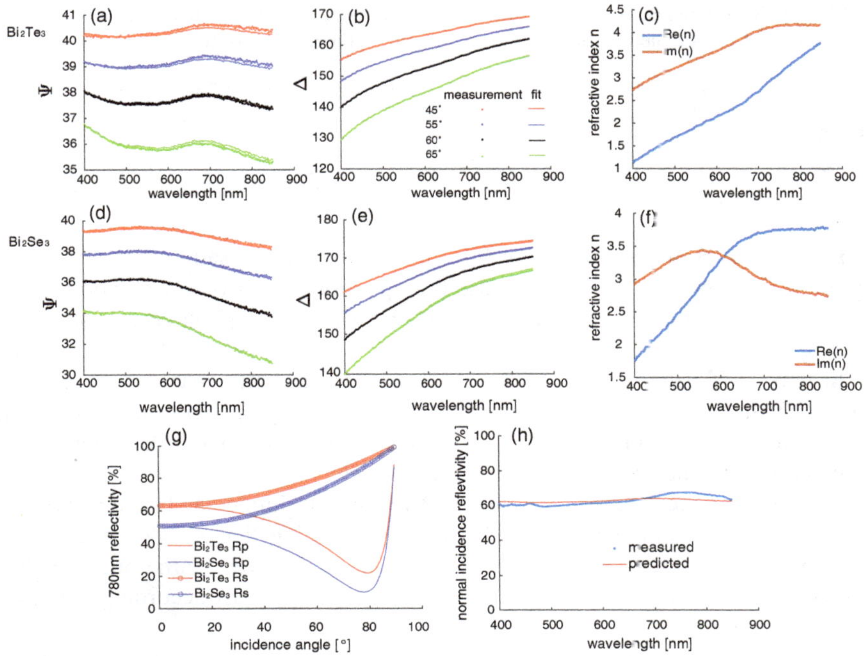

Fig. 7.11 Ellipsometry data for Bi_2Te_3 and Bi_2Se_3. In (a) and (b) for Bi_2Te_3, we plot Ψ and Δ from measurement (dots) and fit result (solid curve). The reflectivity measurement was conducted under four different incidence angles. The fit uses a multilayer reflection model and takes into account the thickness of film and the refractive index of the substrate. The complex refractive index is shown in (c). (d), (e) and (f) Bi_2Se_3 ellipsometry results. (g) The s and p reflectivity of 780 nm light as a function of incidence angle. In (h) we show the reflectivity of a 250 nm thick Bi2Te3 measured under normal incidence (with Filmtrics) and the predicted reflectivity

Filmetrics F40) and the reflectivity predicted from the measured refractive indices, showing good agreement.

Theory Calculations

DFT calculations are performed in the local density approximation (LDA) [51] employing the Hartwigsen-Goedecker-Hutter (HGH) norm-conserving pseudopotentials [52] using the ABINIT code [53, 54]. Fully relativistic corrections and spin-orbit coupling (SOC) are included in the pseudopotential. For electronic band structure calculations, Brillouin zone (BZ) integrations are performed in a $12 \times 12 \times 1$ Monkhorst-Pack (MP) **k**-points mesh in the slab calculations and $8\times8\times8$ mesh in the bulk calculations. An energy cutoff for the plane waves of 15 Ha is used. For bulk phonon calculations using density functional perturbation theory, a $12\times12\times12$

Table 7.3 Bi_2Te_3 and Bi_2Se_3 lattice parameters taken from ref. [55]. a, c are the hexagonal lattice constants, and u,v are the internal parameters describing the position of the atoms inside the unit cell

	a	c	δ_1	δ_2
Bi_2Te_3	4.386 Å	30.497 Å	0.2095	0.4000
Bi_2Se_3	4.143 Å	28.636 Å	0.2117	0.4008

and 6×6×6 **k**-points and q-points MP meshes are used to sample electronic and vibrational states, respectively. We use bulk phonons in all our calculations since they give an accurate representation of the zone center A_{1g} phonons in a slab with many quintuple layers.

In bulk DFT calculations, we use the experimental lattice parameters given in Table 7.3. In the slab calculations, we fully relax the atomic positions along the [111] direction, i.e., direction perpendicular to the surface, while keeping the in-plane lattice parameters as the experimental values.

We calculate the forces in the photoexcited system immediately after photoexcitation using constrained DFT (cDFT) [56], assuming that n valence electrons are promoted to conduction band. In cDFT calculation, we assume one chemical potential, i.e., electron and hole populations thermalize rapidly according to Fermi-Dirac distribution.

For the photoexcited system, with electron-hole carrier density n, we obtain atomic forces F_i in the z or [111] direction on each atom i in the unit cell as:

$$F_i(n) = F_i^0(n) + \sum_{j=1}^{5}[z_j - z_j^0]\Phi_{ij}(n). \tag{7.16}$$

The interatomic force matrix $\Phi_{ij}(n)$ depends on the carrier density and gives the phonon frequencies $\omega_\lambda(n)$ in the photoexcited system. Note that the atomic forces $F_i^0(n)$ are the forces obtained in the photoexcited system for the equilibrium atomic positions z_j^0 in the ground state. The equilibrium atomic positions of the photoexcited system depend on n and can be found from:

$$z_i^{equil}(n) = z_i^0 - \sum_{j=1}^{5}\Phi_{ij}^{-1}(n)F_j^0(n). \tag{7.17}$$

Neglecting any phonon damping, the amplitude A_i of atomic motion immediately after photoexcitation can be calculated from $z_i^0 - z_i^{equil}(n)$.

Figure 7.12 shows some of the results of cDFT simulations. The atoms upon photoexcitation shift towards coordinates of possible higher symmetry (i.e., fractional coordinates move towards integer multiples of 1/5), which is consistent with the experimental observation of atomic motion direction suggested by the diffraction intensity change in Fig. 7.13. The simulations reproduce the atomic

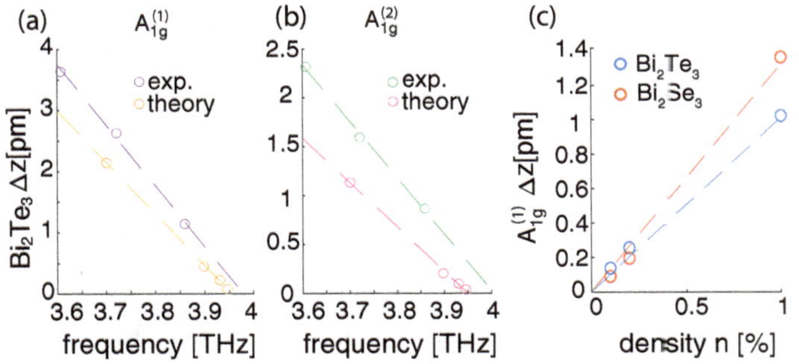

Fig. 7.12 (a) The constrained DFT simulations of photoexcited Bi_2Te_3 atomic displacements for $A_{1g}^{(1)}$ as a function of $A_{1g}^{(2)}$ frequency. (b) Same as (a) but for $A_{1g}^{(2)}$ mode atomic displacements. (c) cDFT simulation shows that Bi_2Te_3 and Bi_2Se_3 are driven at similar amplitudes given the same excitation density

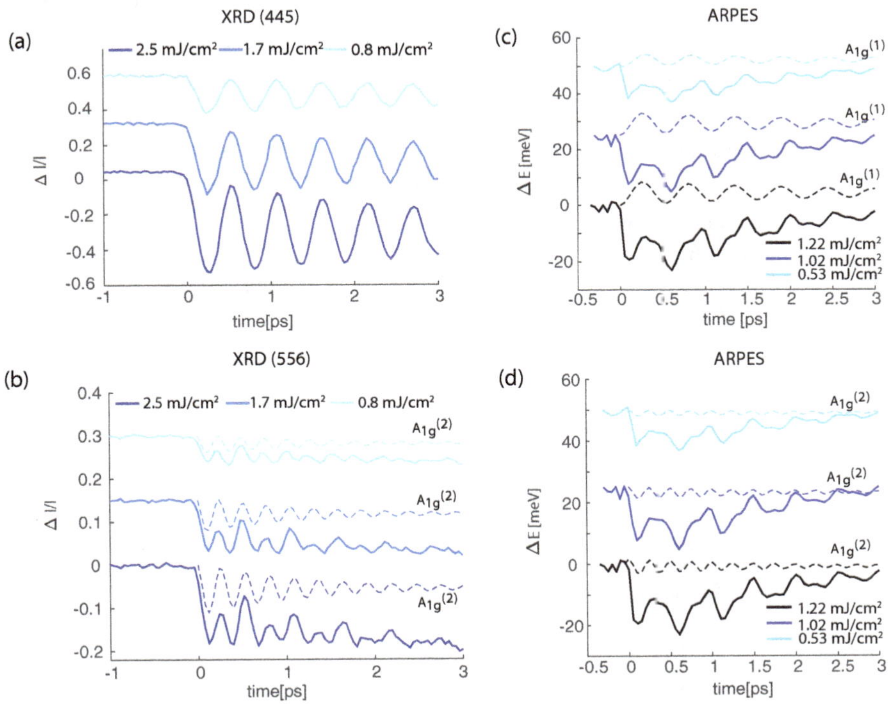

Fig. 7.13 The initial phase displacive excitation of coherent phonons. (a) XRD on peak (445) featuring only mode $A_{1g}^{(1)}$. (b) XRD of peak (556), which contains the $A_{1g}^{(2)}$ component in broken lines. Both A_{1g} modes initially start from the extremum of a sinusoidal function, indicating a well-defined phase (approximately 0 or π)

motion amplitude of A_{1g} modes as a function of the $A_{1g}^{(2)}$ frequency shift. They also correctly predict that Bi_2Te_3 and Bi_2Se_3 are driven at similar amplitudes given the same excitation density, consistent with the experimental results in Fig. 7.2(b).

References

1. Y. Huang, J.D. Querales-Flores, S.W. Teitelbaum, J. Cao, T. Henighan, H. Liu, M. Jiang, G. De la Peña, V. Krapivin, J. Haber, T. Sato, M. Chollet, D. Zhu, T. Katayama, R. Power, M. Allen, C.R. Rotundu, T.P. Bailey, C. Uher, M. Trigo, P.S. Kirchmann, É.D. Murray, Z.-X. Shen, I. Savic, S. Fahy, J.A. Sobota, D.A. Reis, Ultrafast measurements of mode-specific deformation potentials of Bi_2Te_3 and Bi_2Se_3. Phys. Rev. X **13**(4), 041050 (2023). https://doi.org/10.1103/PhysRevX.13.041050
2. D.-X. Qu, Y.S. Hor, J. Xiong, R.J. Cava, N.P. Ong, Quantum oscillations and hall anomaly of surface states in the topological insulator Bi_2Te_3. Science **329**(5993), 821–824 (2010)
3. J.G. Analytis, R.D. McDonald, S.C. Riggs, J.-H. Chu, G.S. Boebinger, I.R. Fisher, Two-dimensional surface state in the quantum limit of a topological insulator. Nat. Phys. **6**(12), 960–964 (2010)
4. F. Xiu, L. He, Y. Wang, L. Cheng, L.-T. Chang, M. Lang, G. Huang, X. Kou, Y. Zhou, X. Jiang et al., Manipulating surface states in topological insulator nanoribbons. Nat. Nanotechnol. **6**(4), 216–221 (2011)
5. B. Hamdou, J. Gooth, A. Dorn, E. Pippel, K. Nielsch, Surface state dominated transport in topological insulator Bi_2Te_3 nanowires. Appl. Phys. Lett. **103**(19), 193107 (2013)
6. K. Hoefer, C. Becker, D. Rata, J. Swanson, P. Thalmeier, L.H. Tjeng, Intrinsic conduction through topological surface states of insulating Bi_2Te_3 epitaxial thin films. Proc. Natl. Acad. Sci. **111**(42), 14979–14984 (2014)
7. R.C. Hatch, M. Bianchi, D. Guan, S. Bao, J. Mi, B.B. Iversen, L. Nilsson, L. Hornekær, P. Hofmann, Stability of the Bi_2Se_3 (111) topological state: electron-phonon and electron-defect scattering. Phys. Rev. B **83**(24), 241303 (2011)
8. T. Kondo, Y. Nakashima, Y. Ota, Y. Ishida, W. Malaeb, K. Okazaki, S. Shin, M. Kriener, S. Sasaki, K. Segawa et al., Anomalous dressing of dirac fermions in the topological surface state of Bi_2Se_3, Bi_2Te_3, and Cu-doped Bi_2Se_3. Phys. Rev. Lett. **110**(21), 217601 (2013)
9. C. Chen, Z. Xie, Y. Feng, H. Yi, A. Liang, S. He, D. Mou, J. He, Xu. Y. Peng, X. Liu, et al., Tunable dirac fermion dynamics in topological insulators. Sci. Rep. **3**(1), 1–6 (2013)
10. Z.-H. Pan, A.V. Fedorov, D. Gardner, Y.S. Lee, S. Chu, T. Valla, Measurement of an exceptionally weak electron-phonon coupling on the surface of the topological insulator Bi_2Se_3 using angle-resolved photoemission spectroscopy. Phys. Rev. Lett. **108**, 187001 (2012)
11. R. Heid, I.Y. Sklyadneva, E.V. Chulkov, Electron-phonon coupling in topological surface states: the role of polar optical modes. Sci. Rep. **7**(1), 1–9 (2017)
12. A. Ruckhofer, D. Campi, M. Bremholm, P. Hofmann, G. Benedek, M. Bernasconi, W.E. Ernst, A. Tamtögl, Terahertz surface modes and electron-phonon coupling on $Bi_2Se_3(111)$. Phys. Rev. Res. **2**(2), 023186 (2020)
13. A. Tamtögl, P. Kraus, N. Avidor, M. Bremholm, E.M.J. Hedegaard, B.B. Iversen, M. Bianchi, P. Hofmann, J. Ellis, W. Allison et al., Electron-phonon coupling and surface debye temperature of $Bi_2Te_3(111)$ from helium atom scattering. Phys. Rev. B **95**(19), 195401 (2017)
14. X. Zhu, L. Santos, C. Howard, R. Sankar, F.C. Chou, C. Chamon, M. El-Batanouny, Electron-phonon coupling on the surface of the topological insulator Bi_2Se_3 determined from surface-phonon dispersion measurements. Phys. Rev. Lett. **108**(18), 185501 (2012)
15. F. Liang, E. Berg, Odd-parity topological superconductors: theory and application to $cu_x bi_2 se_3$. Phys. Rev. Lett. **105**, 097001 (2010)

16. F. Liang, Odd-parity topological superconductor with nematic order: application to $cu_x bi_2 se_3$. Phys. Rev. B **90**, 100509 (2014)
17. K. Matano, M. Kriener, K. Segawa, Y. Ando, G.-Q. Zheng, Spin-rotation symmetry breaking in the superconducting state of $Cu_xBi_2Se_3$. Nat. Phys. **12**(9), 852–854 (2016)
18. T. Asaba, B.J. Lawson, C. Tinsman, L. Chen, P. Corbae, G. Li, Y. Qiu, Y.S. Hor, L. Fu, L. Li, Rotational symmetry breaking in a trigonal superconductor nb-doped Bi_2Se_3. Phys. Rev. X **7**(1), 011009 (2017)
19. S. Yonezawa, K. Tajiri, S. Nakata, Y. Nagai, Z. Wang, K. Segawa, Y. Ando, Y. Maeno, Thermodynamic evidence for nematic superconductivity in $Cu_xBi_2Se_3$. Nat. Phys. **13**(2), 123–126 (2017)
20. K. Willa, R. Willa, K.W. Song, G.D. Gu, J.A. Schneeloch, R. Zhong, A.E. Koshelev, W.-K. Kwok, U. Welp, Nanocalorimetric evidence for nematic superconductivity in the doped topological insulator $Sr_{0.1}Bi_2Se_3$. Phys. Rev. B **98**(18), 184509 (2018)
21. J. Wang, K. Ran, S. Li, Z. Ma, S. Bao, Z. Cai, Y. Zhang, K. Nakajima, S. Ohira-Kawamura, P. Čermák et al., Evidence for singular-phonon-induced nematic superconductivity in a topological superconductor candidate sr0. $1Bi_2Se_3$. Nat. Commun. **10**(1), 1–6 (2019)
22. X. Li, T. Qiu, J. Zhang, E. Baldini, L. Jian, A.M. Rappe, K.A. Nelson, Terahertz field–induced ferroelectricity in quantum paraelectric SrTiO3. Science **364**(6445), 1079–1082 (2019)
23. J. Xiao, L. Kou, C.Y. Yam, T. Frauenheim, B. Yan, Toward rational design of catalysts supported on a topological insulator substrate. ACS Cataly. **5**(12), 7063–7067 (2015)
24. G.J. Kroes, Frontiers in surface scattering simulations. Science **321**(5890), 794–797 (2008)
25. Y. Tokura, Photoinduced phase transition: a tool for generating a hidden state of matter. J. Phys. Soc. Jpn. **75**(1), 011001–011001 (2006)
26. D.N. Basov, R.D. Averitt, D. Hsieh, Towards properties on demand in quantum materials. Nat. Mater. **16**(11), 1077–1088 (2017)
27. L. Luo, X. Yang, X. Liu, Z. Liu, C. Vaswani, D. Cheng, M. Mootz, X. Zhao, Y. Yao, C.-Z. Wang et al., Ultrafast manipulation of topologically enhanced surface transport driven by mid-infrared and terahertz pulses in Bi_2Se_3. Nat. Commun. **10**(1), 1–9 (2019)
28. E.J. Sie, C.M. Nyby, C.D. Pemmaraju, S.J. Park, X. Shen, J. Yang, M.C. Hoffmann, B.K. Ofori-Okai, R. Li, A.H. Reid et al., An ultrafast symmetry switch in a weyl semimetal. Nature **565**(7737), 61–66 (2019)
29. C. Vaswani, L.-L. Wang, D.H. Mudiyanselage, Q. Li, P.M. Lozano, G.D. Gu, D. Cheng, B. Song, L. Luo, R.H.J. Kim, C. Huang, Z. Liu, M. Mootz, I.E. Perakis, Y. Yao, K.M. Ho, J. Wang, Light-driven raman coherence as a nonthermal route to ultrafast topology switching in a dirac semimetal. Phys. Rev. X **10**, 021013 (2020)
30. N. Aryal, X. Jin, Q. Li, A.M. Tsvelik, W. Yin, Topological phase transition and phonon-space dirac topology surfaces in ZrTe 5. Phys. Rev. Lett. **126**(1), 016401 (2021)
31. L. Luo, D. Cheng, B. Song, L.L. Wang, C. Vaswani, P.M. Lozano, G. Gu, C. Huang, R.H.J. Kim, Z. Liu, J.M. Park, Y. Yao, K. Ho, I.E. Perakis, Q. Li, J. Wang, A light-induced phononic symmetry switch and giant dissipationless topological photocurrent in ZrTe5. Nat. Mater. **20**(3), 329–334 (2021)
32. J. Bardeen, W. Shockley, Deformation potentials and mobilities in non-polar crystals. Phys. Rev. **80**, 72–80 (1950)
33. S. Gerber, S.-L. Yang, D. Zhu, H. Soifer, J.A. Sobota, S. Rebec, J.J. Lee, T. Jia, B. Moritz, C. Jia et al., Femtosecond electron-phonon lock-in by photoemission and X-ray free-electron laser. Science **357**(6346), 71–75 (2017)
34. M. Na, A.K. Mills, F. Boschini, M. Michiardi, B. Nosarzewski, R.P. Day, E. Razzoli, A. Sheyerman, M. Schneider, G. Levy et al., Direct determination of mode-projected electron-phonon coupling in the time domain. Science **366**(6470), 1231–1236 (2019)
35. H. Xiong, J.A. Sobota, S.-L. Yang, H. Soifer, A. Gauthier, M.-H. Lu, Y.-Y. Lv, S.-H. Yao, D. Lu, M. Hashimoto, P.S. Kirchmann, Y.-F. Chen, Z.-X. Shen, Three-dimensional nature of the band structure of $ZrTe_5$ measured by high-momentum-resolution photoemission spectroscopy. Phys. Rev. B **95**, 195119 (2017)

36. J.A. Sobota, S.-L. Yang, D. Leuenberger, A.F. Kemper, J.G. Analytis, I.R. Fisher, P.S. Kirchmann, T.P. Devereaux, Z.-X. Shen, Distinguishing bulk and surface electron-phonon coupling in the topological insulator Bi_2Se_3 using time-resolved photoemission spectroscopy. Phys. Rev. Lett. **113**, 157401 (2014)
37. J.A. Sobota, S.W. Teitelbaum, Y. Huang, J.D. Querales-Flores, R. Power, M. Allen, C.R. Rotundu, T.P. Bailey, C. Uher, T. Henighan, M. Jiang, D. Zhu, M. Chollet, T. Sato, M. Trigo, É.D. Murray, I. Savić, P.S. Kirchmann, S. Fahy, D.A. Reis, Z.-X. Shen, Influence of local symmetry on lattice dynamics coupled to topological surface states. Phys. Rev. B **107**, 014305 (2023)
38. A. Pertsova, C.M. Canali, Probing the wavefunction of the surface states in Bi2Se3 topological insulator: a realistic tight-binding approach. New J. Phys. **16**(6), 063022 (2014)
39. H.J. Zeiger, J. Vidal, T.K. Cheng, E.P. Ippen, G. Dresselhaus, M.S. Dresselhaus, Theory for displacive excitation of coherent phonons. Phys. Rev. B **45**, 768–778 (1992)
40. H.-W. Jeon, H.-P. Ha, D.-B. Hyun, J.-D. Shim, Electrical and thermoelectrical properties of undoped Bi_2Te_3-Sb_2Te_3 and Bi_2Te_3-Sb_2Te_3-Sb2Se3 single crystals. J. Phys. Chem. Solids **52**(4), 579–585 (1991)
41. Y.-M. Sheu, Y.J. Chien, C. Uher, S. Fahy, D.A. Reis, Free-carrier relaxation and lattice heating in photoexcited bismuth. Phys. Rev. B **87**(7), 075429 (2013)
42. D. Boschetto, E.G. Gamaly, A.V. Rode, B. Luther-Davies, D. Glijer, T. Garl, O. Albert, A. Rousse, J. Etchepare, Small atomic displacements recorded in bismuth by the optical reflectivity of femtosecond laser-pulse excitations. Phys. Rev. Lett. **100**(2), 027404 (2008)
43. A.Q. Wu, X. Xu, Coupling of ultrafast laser energy to coherent phonons in bismuth. Appl. Phys. Lett. **90**(25), 251111 (2007)
44. O.V. Misochko, M. Hase, K. Ishioka, M. Kitajima, Observation of an amplitude collapse and revival of chirped coherent phonons in bismuth. Phys. Rev. Lett. **92**(19), 197401 (2004)
45. M.F. DeCamp, D.A. Reis, P.H. Bucksbaum, R. Merlin, Dynamics and coherent control of high-amplitude optical phonons in bismuth. Phys. Rev. B **64**(9), 092301 (2001)
46. C. B. Satterthwaite, R. W. Ure, Electrical and thermal properties of bi_2 te_3. Phys. Rev. **108**, 1164–1170 (1957)
47. D. H. Auston, C. V. Shank, Picosecond ellipsometry of transient electron-hole plasmas in germanium. Phys. Rev. Lett. **32**, 1120–1123 (1974)
48. J. F. Young, H. M. van Driel, Ambipolar diffusion of high-density electrons and holes in ge, si, and gaas: Many-body effects. Phys. Rev. B. **26**, 2147–2158 (1982)
49. P. Janícek, C. Drašar, L. Benesš, P. Lošták, Thermoelectric properties of Tl-doped Bi_2Se_3 single crystals. Cryst. Res. Technol. **44**(5), 505–510 (2009)
50. T. Förster, P. Krüger, M. Rohlfing, gw calculations for Bi_2Te_3 and Sb_2Te_3 thin films: electronic and topological properties. Phys. Rev. B **93**, 205442 (2016)
51. J.P. Perdew, A. Zunger, Self-interaction correction to density-functional approximations for many-electron systems. Phys. Rev. B **23**, 5048–5079 (1981)
52. C. Hartwigsen, S. Goedecker, J. Hutter, Relativistic separable dual-space gaussian pseudopotentials from H to Rn. Phys. Rev. B **58**, 3641–3662 (1998)
53. X. Gonze, B. Amadon, P.-M. Anglade, J.-M. Beuken, F. Bottin, P. Boulanger, F. Bruneval, D. Caliste, R. Caracas, M. Côté, T. Deutsch, L. Genovese, P. Ghosez, M. Giantomassi, S. Goedecker, D.R. Hamann, P. Hermet, F. Jollet, G. Jomard, S. Leroux, M. Mancini, S. Mazevet, M.J.T. Oliveira, G. Onida, Y. Pouillon, T. Rangel, G.-M. Rignanese, D. Sangalli, R. Shaltaf, M. Torrent, M.J. Verstraete, G. Zerah, J.W. Zwanziger, Abinit: first-principles approach to material and nanosystem properties. Comput. Phys. Commun. **180**(12), 2582–2615 (2009). 40 Years of CPC: A celebratory issue focused on quality software for high performance, grid and novel computing architectures
54. X. Gonze, F. Jollet, F. Abreu Araujo, D. Adams, B. Amadon, T. Applencourt, C. Audouze, J.-M. Beuken, J. Bieder, A. Bokhanchuk, E. Bousquet, F. Bruneval, D. Caliste, M. Côté, F. Dahm, F. Da Pieve, M. Delaveau, M. Di Gennaro, B. Dorado, C. Espejo, G. Geneste, L. Genovese, A. Gerossier, M. Giantomassi, Y. Gillet, D.R. Hamann, L. He, G. Jomard, J. Laflamme Janssen, S. Le Roux, A. Levitt, A. Lherbier, F. Liu, I. Lukačević, A. Martin, C. Martins, M.J.T. Oliveira,

S. Poncé, Y. Pouillon, T. Rangel, G.-M. Rignanese, A.H. Romero, B. Rousseau, O. Rubel, A.A. Shukri, M. Stankovski, M. Torrent, M.J. Van Setten, B. Van Troeye, M.J. Verstraete, D. Waroquiers, J. Wiktor, B. Xu, A. Zhou, J.W. Zwanziger, Recent developments in the abinit software package. Comput. Phys. Commun. **205**, 106–131 (2016)

55. S. Nakajima, The crystal structure of $Bi_2Te_{3-x}Se_x$. J. Phys. Chem. Solids **24**(3), 479–485 (1963)
56. É.D. Murray, D.M. Fritz, J.K. Wahlstrand, S. Fahy, D.A. Reis, Effect of lattice anharmonicity on high-amplitude phonon dynamics in photoexcited bismuth. Phys. Rev. B **72**, 060301 (2005)

Chapter 8
Outlook

Abstract The continuous development of free-electron lasers (FELs) holds great promise for advancing experimental capabilities in various ways. Anticipated progress includes improvements in pulse duration, coherence properties, and spectral and energy stability, as well as increased repetition rates. These advancements will not only facilitate the integration of X-ray probes developed at synchrotron facilities into the FEL toolbox but also enable new techniques unique to FELs, leading to the emergence of new physics. The specific problems addressed in the main results of this thesis, such as the quantitative measurement of electron-phonon coupling, the interplay between electron orbitals, interatomic interactions, lattice instability, and the distinctions between nonequilibrium and thermal states, will be revisited in the context of future FEL development.

Over the past decade, there has been a remarkable surge in the investigation of photoinduced emergent phenomena. However, achieving a comprehensive and universally applicable understanding of the microscopic mechanisms underlying these phenomena remains a significant challenge. While theoretical approaches can offer valuable insights, the calculation of nonequilibrium states is still in its nascent stages, primarily due to the overwhelming complexity involved. Consequently, most theoretical calculations in this thesis focus on equilibrium states.

On the experimental front, the development of free-electron lasers (FELs) with enhanced capabilities, including high repetition rates, attosecond time resolution, and advanced end-stations, holds great promise for addressing crucial questions pertaining to the nonequilibrium dynamics of solid materials. It is anticipated that significant advancements in experimental capabilities, with an order of magnitude improvement, will lead to the emergence of new physics and further deepen our understanding of these phenomena.

Within the next five years, LCLS will fully upgrade to 1MHz repetition rate operating from soft X-ray to hard X-ray, and will also add new end-stations such as qRIXS [1], which is ideal for resonant inelastic X-ray scattering (RIXS) as well as X-ray photon correlation spectroscopy (XPCS) (the split-and-delay setup that serves the purpose of the latter has been commissioned [2, 3]). RIXS involves measuring

Y. Huang, *Towards the Optical Control of Resonantly Bonded Materials*, Springer Theses, https://doi.org/10.1007/978-3-031-42826-5_8

the X-ray energy loss at the core-hole resonance, as a function of momentum transfer, in order to characterize collective electronic excitations and gain insights into short-range correlations. It is effectively an inelastic scattering experiment discussed in Chap. 2 but is sensitive to spin, charge, or orbital excitations rather than phonons. Time-resolved RIXS has shown its power in measuring magnetic excitations [4, 5] and diffuse charge density wave order [6]. RIXS will greatly benefit from the high-flux narrowband X-rays produced by an FEL. XPCS, on the other hand, uses low-intensity FEL pulses so as to avoid creating extra fluctuations in the sample and requires a highly coherent beam as well as instrument stability. High repetition rate FEL will enhance the feasibility of using statistics to extract photon correlation [7] in a XPCS measurement.

LCLS also runs the XLEAP program which targets producing attosecond X-ray pulses that are more than a million times brighter than the attosecond pulses of similar energy ever produced. Its first commissioning happened in 2019 [8]. As discussed in Chap. 2, the nonequilibrium dynamics of crystals could, in principle, be measured either in the frequency domain or the time domain. The time resolution of an optical pump X-ray probe setup (e.g., the XPP end-station at LCLS) is roughly 50fs. It is too slow to directly probe the electron dynamics in the time domain (the practical reason that we have been studying phonons). In the future, attosecond capability at LCLS will enable time-domain probing of attosecond-timescale dynamics [9], potentially in solids. Dixit et al. had several good references [10–12] discussing the probing of electron density-density fluctuations in the far-field regime and in the time domain.

In Chap. 5, the influence of the pump photon energy on the population of orbitals and the resulting structural preferences was demonstrated. This highlights the potential for tailored pump pulses to prepare the excited states and explore various aspects of material dynamics. I give several examples below. The duration of the pump pulse can be adjusted, allowing control over the timing of quenching processes, as demonstrated in studies on quenching dynamics [13, 14]. Additionally, using light with exotic polarization states enables the breaking of selective symmetries [15]. By using mid-infrared (MIR) and terahertz (THz) pumps, one can deposit more energy into the lower-energy excitations than above-bandgap electronic excitations, and by this means intriguing metastable phases have been discovered [16, 17]. Moreover, by creating standing waves within materials with transient grating techniques, we can investigate charge, spin, and energy transfer processes at atomic scales [18].

An important direction for future research lies in integrating less conventional pump sources with recently developed probing techniques. Specifically, the combination of terahertz (THz) and mid-infrared (MIR) excitation with time-resolved diffuse scattering probes, such as Fourier transform inelastic X-ray scattering (FTIXS), shows great promise in advancing our understanding of novel nonequilibrium phenomena. One area of interest is the exploration and verification of the concept of nonlinear phononics, which refers to the manipulation of lattice vibrations and their coupling to other excitations using THz or MIR light sources. There are few reports up to date on nonlinear phononics at finite $\mathbf{q}$, and FTIXS can nicely fill this gap [19].

In Chap. 6, the simultaneous study of valence charge dynamics and lattice motion is motivated, aiming to gain a comprehensive understanding of material behavior. While there are stable and high repetition rate tabletop sources available for probing attosecond electron dynamics spectroscopically, the ability to image electrons in real space would be a significant advancement. Several references, including Yavas et al. [20], Leedahl et al. [21], and Amorese et al. [22], demonstrate direct orbital imaging based on inelastic X-ray scattering, which could be extended to an FEL end-station. Alternatively, Glover et al. [23] propose a scheme for probing the spatial distribution of optically induced charges using nonlinear X-ray optical mixing signals, which could also be adapted to investigate nonequilibrium charge distributions induced by UV or soft X-ray pumps.

The time-resolved diffuse scattering technique, also known as FTIXS, when performed on resonance, can be considered a time-domain version of RIXS (and not to be confused with time-resolved RIXS). Such technique is currently challenging due to the tiny cross section of resonant diffraction and the even weaker resonant diffuse scattering. However, with the future high repetition rate machine, this becomes a promising avenue to explore.

In Chap. 7, it was demonstrated that multimodal experiments can suffer from certain limitations, including differences in sample preparation and variations in the pump laser setup, in addition to the inherent differences between the probing techniques themselves. However, the integration of angle-resolved photoemission spectroscopy (ARPES) and diffraction in a simultaneous experiment can be realized at an FEL facility. Employing higher photon energies (such as hard X-rays) for photoemission has additional advantages. In hard X-ray, the penetration depth is sensitive to the change in incidence angles, which allows for the differentiation of the electronic state response in the bulk of the material from that of surface states, simply by varying the incidence angle of the laser beam in an ARPES measurement. Combining ARPES and diffraction techniques in a well-controlled and coordinated manner opens up avenues for quantitatively studying the interplay between electronic and lattice degrees of freedom, ultimately enhancing our understanding of fundamental physics in condense matter.

Spectroscopists always find that they need to balance between exploring new techniques using well-understood materials and exploring new materials with well-understood techniques. But we have welcomed an era of FEL when one can be comfortable taking either approach.

References

1. M. Dunne. Lcls strategic facility development plan, in *SLAC National Accelerator Laboratory, Menlo Park, CA, USA* (2020), p. 55
2. Y. Sun, N. Wang, S. Song, P. Sun, M. Chollet, T. Sato, T.B. van Driel, S. Nelson, R. Plumley, J. Montana-Lopez et al., Compact hard X-ray split-delay system based on variable-gap channel-cut crystals. Opt. Lett. **44**(10), 2582–2585 (2019)

3. H. Li, Y. Sun, J. Vila-Comamala, T. Sato, S. Song, P. Sun, M.H. Seaberg, N. Wang, J.B. Hastings, M. Dunne et al., Generation of highly mutually coherent hard-X-ray pulse pairs with an amplitude-splitting delay line. Phys. Rev. Res. **3**(4), 043050 (2021)
4. M.P.M. Dean, Y. Cao, X. Liu, S. Wall, D. Zhu, R. Mankowsky, V. Thampy, X.M. Chen, J.G. Vale, D. Casa, J. Kim, A.H. Said, P. Juhas, R. Alonso-Mori, J.M. Glownia, A. Robert, J. Robinson, M. Sikorski, S. Song, M. Kozina, H. Lemke, L. Patthey, S. Owada, T. Katayama, M. Yabashi, Y. Tanaka, T. Togashi, J. Liu, C. Rayan Serrao, B.J. Kim, L. Huber, C.L. Chang, D.F. McMorrow, M. Forst, J.P. Hill. Ultrafast energy- and momentum-resolved dynamics of magnetic correlations in the photo-doped Mott insulator Sr2IrO4. Nat. Mater. **15**(6), 601–605 (2016)
5. D.G. Mazzone, D. Meyers, Y. Cao, J.G. Vale, C.D. Dashwood, Y. Shi, A.J.A. James, N.J. Robinson, J. Lin, V. Thampy, Y. Tanaka, A.S. Johnson, H. Miao, R. Wang, T.A. Assefa, J. Kim, D. Casa, R. Mankowsky, D. Zhu, R. Alonso-Mori, S. Song, H. Yavas, T. Katayama, M. Yabashi, Y. Kubota, S. Owada, J. Liu, J. Yang, R.M. Konik, I.K. Robinson, J.P. Hill, D.F. McMorrow, M. Forst, S. Wall, X. Liu, M.P.M. Dean, Laser-induced transient magnons in Sr3Ir2O7 throughout the Brillouin zone. Proc. Natl. Acad. Sci. U. S. A. **118**(22), e2103696118 (2021)
6. M. Mitrano, S. Lee, A.A. Husain, L. Delacretaz, M. Zhu, G. De La Peña Munoz, S.X.L. Sun, Y.I. Joe, A.H. Reid, S.F. Wandel, G. Coslovich, W. Schlotter, T. Van Driel, J. Schneeloch, G.D. Gu, S. Hartnoll, N. Goldenfeld, P. Abbamonte, Ultrafast time-resolved x-ray scattering reveals diffusive charge order dynamics in $La_{2-x}Ba_xCuO_4$. Sci. Adv. **5**(8) (2019).
7. S.O. Hruszkewycz, M. Sutton, P.H. Fuoss, B. Adams, S. Rosenkranz, K.F. Ludwig, W. Roseker, D. Fritz, M. Cammarata, D. Zhu, S. Lee, H. Lemke, C. Gutt, A. Robert, G. Grübel, G.B. Stephenson, High contrast X-ray speckle from atomic-scale order in liquids and glasses. Phys. Rev. Lett. **109**(18), 1–5 (2012)
8. J. Duris, S. Li, T. Driver, E.G. Champenois, J.P. MacArthur, A.A. Lutman, Z. Zhang, P. Rosenberger, J.W. Aldrich, R. Coffee et al., Tunable isolated attosecond x-ray pulses with gigawatt peak power from a free-electron laser. Nat. Photon. **14**(1), 30–36 (2020)
9. J.T. O'Neal, E.G. Champenois, S. Oberli, R. Obaid, A. Al-Haddad, J. Barnard, N. Berrah, R. Coffee, J. Duris, G. Galinis, D. Garratt, J.M. Glownia, D. Haxton, P. Ho, S. Li, X. Li, J. MacArthur, J.P. Marangos, A. Natan, N. Shivaram, D.S. Slaughter, P. Walter, S. Wandel, L. Young, C. Bostedt, P.H. Bucksbaum, A. Picón, A. Marinelli, J.P. Cryan, Electronic population transfer via impulsive stimulated X-ray Raman scattering with attosecond soft-X-ray pulses. Phys. Rev. Lett. **125**, 073203 (2020)
10. G. Dixit, O. Vendrell, R. Santra, Imaging electronic quantum motion with light. Proc. Natl. Acad. Sci. **109**(29), 11636–11640 (2012)
11. G. Dixit, J.M. Slowik, R. Santra. Theory of time-resolved nonresonant X-ray scattering for imaging ultrafast coherent electron motion. Phys. Rev. A **89**(4), 043409 (2014)
12. G. Dixit, J.M. Slowik, R. Santra, Proposed imaging of the ultrafast electronic motion in samples using X-ray phase contrast. Phys. Rev. Lett. **110**(13), 137403 (2013)
13. W.H. Zurek, U. Dorner, P. Zoller, Dynamics of a quantum phase transition. Phys. Rev. Lett. **95**(10), 105701 (2005)
14. H. Bernien, S. Schwartz, A. Keesling, H. Levine, A. Omran, H. Pichler, S. Choi, A.S. Zibrov, M. Endres, M. Greiner, V. Vuletic, M.D. Lukin, Probing many-body dynamics on a 51-atom quantum simulator. Nature **551**(7682), 579–584 (2017)
15. T.V. Trevisan, P.V. Arribi, O. Heinonen, R.-J. Slager, P.P. Orth, Bicircular light floquet engineering of magnetic symmetry and topology and its application to the dirac semimetal cd_3as_2. Phys. Rev. Lett. **128**, 066602 (2022)
16. X. Li, T. Qiu, J. Zhang, E. Baldini, L. Jian, A.M. Rappe, K.A. Nelson, Terahertz field–induced ferroelectricity in quantum paraelectric SrTiO3. Science **364**(6445), 1079–1082 (2019)
17. T.F. Nova, A.S. Disa, M. Fechner, A. Cavalleri, Metastable ferroelectricity in optically strained SrTiO3. Science **364**(6445), 1075–1079 (2019)

18. U. Wagner, G. Knopp, M. Beck, J. Krempasky, S. Zerdane, R. Follath, P. Beaud, I. Mochi, M. Chergui, C. Cirelli, G. Ingold, A. Kubec, M. Lebugle, G.F. Mancini, V. Esposito, B. Rösner, C. Pradervand, C. Milne, B. Pedrini, U. Flechsig, C. Svetina, R. Mankowsky, J. Rouxel, C. David, F. Koch, H.T. Lemke, G. Seniutinas. Towards X-ray transient grating spectroscopy. Opt. Lett. **44**(3), 574–577 (2019)
19. S.W. Teitelbaum, T. Henighan, Y. Huang, H. Liu, M.P. Jiang, D. Zhu, M. Chollet, T. Sato, É.D. Murray, S. Fahy, S. O'Mahony, T.P. Bailey, C. Uher, M. Trigo, D.A. Reis, Direct measurement of anharmonic decay channels of a coherent phonon. Phys. Rev. Lett. **121**, 125901 (2018)
20. H. Yavas, M. Sundermann, K. Chen, A. Amorese, A. Severing, H. Gretarsson, M.W. Haverkort, L.H. Tjeng. Direct imaging of orbitals in quantum materials. Nat. Phys. **15**(6), 559–562 (2019)
21. B. Leedahl, M. Sundermann, A. Amorese, A. Severing, H. Gretarsson, L. Zhang, A.C. Komarek, A. Maignan, M.W. Haverkort, L.H. Tjeng, Origin of Ising magnetism in Ca3Co2O6 unveiled by orbital imaging. Nat. Commun. **10**(1), 1–7 (2019)
22. A. Amorese, B. Leedahl, M. Sundermann, H. Gretarsson, Z. Hu, H.J. Lin, C.T. Chen, M. Schmidt, H. Borrmann, Y. Grin, A. Severing, M.W. Haverkort, L.H. Tjeng, Selective orbital imaging of excited states with X-ray spectroscopy: the example of α -MnS. Phys. Rev. X **11**(1), 1–11 (2021)
23. T.E. Glover, D.M. Fritz, M. Cammarata, T.K. Allison, S. Coh, J.M. Feldkamp, H. Lemke, D. Zhu, Y. Feng, R.N. Coffee, M. Fuchs, S. Ghimire, J. Chen, S. Shwartz, D.A. Reis, S.E. Harris, J.B. Hastings, X-ray and optical wave mixing. Nature **488**(7413), 603–608 (2012)

Printed in the USA
CPSIA information can be obtained
at www.ICGtesting.com
LVHW020059280124
770066LV00003B/263